W0113759

Science, Agriculture, and
the Politics of Research

Also of Interest

† Available in hardcover and paperback.

About the Book and Authors

Science, Agriculture, and
the Politics of Research
Lawrence Busch and William B. Lacy

Although the U.S. agricultural research system is highly esteemed and used as a model for national programs in many developing countries, surprisingly little is known about its operation, which has been publicly funded for nearly a century. This book describes the structure of that system and examines the many factors that influence the choices of research problems by agricultural scientists working within it.

The authors show how the research system has evolved through a complex set of negotiations into its present form. They then examine how individuals, organizations, and disciplines influence scientists' decisions about what to study, as well as their perceptions of research goals and beneficiaries. Drawing throughout on a wealth of primary documents, in-depth interviews, a national survey of over 1,400 practicing scientists, a survey of editors of science journals, and government statistical series, Drs. Busch and Lacy conclude by delineating the key issues that scientists, administrators, and policymakers must address as the public agricultural research system prepares to enter its second century.

Lawrence Busch and William B. Lacy are associate professors of sociology at the University of Kentucky, with appointments in both the College of Agriculture and the College of Arts and Sciences.

Westview Special Studies
in Agriculture Science and Policy

Rural Studies Series
of the Rural Sociological Society

Science, Agriculture, and the Politics of Research

Lawrence Busch and William B. Lacy

Routledge
Taylor & Francis Group

LONDON AND NEW YORK

To our children
Lisa, Rachel, Donovan, and Kristin

First published 1983 by Westview Press

Published 2019 by Routledge
52 Vanderbilt Avenue, New York, NY 10017
2 Park Square, Milton Park, Abingdon, Oxon OX14 4RN

Routledge is an imprint of the Taylor & Francis Group, an informa business

Copyright © 1983 by Taylor & Francis

All rights reserved. No part of this book may be reprinted or reproduced or utilised in any form or by any electronic, mechanical, or other means, now known or hereafter invented, including photocopying and recording, or in any information storage or retrieval system, without permission in writing from the publishers.

Notice:
Product or corporate names may be trademarks or registered trademarks, and are used only for identification and explanation without intent to infringe.

Library of Congress Cataloging in Publication Data
Busch, Lawrence.
    Science, agriculture, and the politics of research.
    (Westview special studies in agriculture science and policy) (Rural studies series, sponsored by the Rural Sociological Society)
    Bibliography: p.
    Includes index.
    1. Agricultural research—United States. 2. Agriculturists—United States. 3. Agricultural research—Government policy—United States. I. Lacy, William B., 1942–   II. Title. III. Series.
S541.B87   1983   630'.72073   82-15923

ISBN 13: 978-0-367-28671-2 (hbk)
ISBN 13: 978-0-367-30217-7 (pbk)

Contents

Tables and Figures

Tables

Figures

Acknowledgments

Many friends, colleagues, and research staff members have directly and indirectly contributed to this book. It is impossible to acknowledge the contribution of each. Still, we would like to recognize several persons as well as institutions that have been particularly helpful. Research funds were provided by the Kentucky Agricultural Experiment Station and by the Ford Foundation. John Myers of the Current Research Information System provided us with a computer tape listing current projects. Carolyn Sachs was extremely helpful in coordinating the mail survey of scientists. Christian Ritter, Lisa Slatin, and Bobbie Sparks assisted in coding the data. Ann Stockham developed the index and also organized the data. Janet Baynham, Sue Lewis, and Greg Taylor aided in the voluminous computer programming and statistical analysis. Rosemary Cheek typed most of the manuscript. Marlene Pettit, Michael Claycomb, Deborah Wheeler, and Penny Hogue also assisted in the typing. Janice Taylor aided in the manuscript typing and ran interference on much of the administrative detail.

Nancy Welt helped us to locate a large number of obscure documents and to assemble the tables in the text. Librarians at the University of Kentucky libraries were especially helpful. In particular, Toni Powell of the Agricultural Library and Sandy Gilchrist of the Margaret I. King Library helped us to unearth various documents.

David Brown, Mike Brazzel, Fred Buttel, James Christenson, C. Milton Coughenour, Reed Hertford, Mark Lancelle, Kenneth Pigg, David Pimentel, Vernon Ruttan, Ann Stockham, Louis Swanson, Sylvan Wittwer, Jane Wolfson, and James Zuiches read significant sections of the manuscript and provided helpful comments. Donald DeLuca pointed out several potential methodological problems. In addition, participants at a workship held at the Ford Foundation in 1980, as well as several members of the Experiment Station Committee on Organization and Policy and the Northeast Experiment Station Directors' Association, provided useful feedback on some of our preliminary research findings. Finally, our wives, Karen and Laura, critiqued and judiciously edited the manuscript and contributed significantly by tolerating us throughout this process.

Of course, the usual disclaimers apply. The interpretation of research findings and recommendations listed herein are not necessarily those of any of the individuals or organizations listed above.

Lawrence Busch
William B. Lacy

Introduction

How to attain organized efficiency, and still retain the individuality, honor the personality, obtain justice for and secure the full welfare of all concerned. This is the great problem of the new epoch.

—Kenyon Butterfield
President, Massachusetts
Agricultural College, 1917

In this work we endeavor to examine factors that influence the choice of research problems by agricultural scientists. Put simply: Why do agricultural scientists do what they do? At first glance it would appear that the answer is obvious. Indeed, when we originally raised the question we were told that "everybody knows that scientists do whatever they can get funding for." In addition, we heard that "they do what's easy" and "they do whatever will produce lots of publications." The "obvious" character of all these replies shows how easily and quickly a plausible rationale is established, but it masks a paucity of research on the process and the products of agricultural research.

The enormous literature on the diffusion of innovations (e.g., Rogers and Shoemaker, 1971) has dealt entirely with the finished products of agricultural research, even though scientific growth may also be viewed as a diffusion process (Crane, 1972). Although a few works have noted the problems confronting agricultural research (e.g., Cheney, 1974; Patterson, 1976), including issues of resource allocation (e.g., Fishel, 1971; Arndt, Dalrymple, and Ruttan, 1977) and productivity (e.g., Lewin, 1972; U.S. Department of Agriculture [USDA], Agricultural Research Service, 1967), there have been only a few brief attempts to examine the social system that permits the creation of valid agricultural research (Jimenez, 1964; Salmon and Hanson, 1964; Dillon, 1976; Fujimoto and Fiske, 1975; Fujimoto and Kopper, 1975). As Albert Moseman (1971:139) put it, "Perhaps the most

1

certain feature about the building of national systems for agricultural research is that neither their significance nor their processes are well understood." It is these aspects of agricultural research that this book seeks to explore.

We believe that such questions must be asked as the system of public agricultural research enters its second century. Food and fiber productivity in the United States is extraordinarily high by both world and historical standards. Nevertheless, doubts have been raised recently about the ability of U.S. agriculture to maintain continued growth in productivity. At the same time, federal funds for agricultural research—in constant dollars—are barely keeping pace with inflation. Some states have maintained their support while others, hard hit by unemployment, have reduced support significantly. The deterioration of financial support for agricultural research is intimately linked to the decline in size and increasing fragmentation of the farm population—fragmentation that is, in part, the result of the research process itself. These problems demand greater understanding of how our public agricultural research system operates, who it serves, and how it will cope with the problems that lie ahead.

On the international scene, despite significant increases in food production over the last several decades, a large segment of the world's population still survives at the brink of starvation. Furthermore, increasing food production and consumption during the coming decades will be more difficult: The productivity of more marginal lands will have to be increased by using relatively low cost inputs. The questions of *access* to land and food by small farmers will have to be better incorporated into the research agenda. Finally, research questions regarding nutrition, marketing, and equitable distribution of food must be addressed.

Since its inception, and particularly since the end of World War II, the U.S. model of agricultural research has been actively diffused throughout the world. Japan, India, Brazil, and other smaller nations have based significant portions of their research system upon the U.S. model. "Institution building" has been a major theme in U.S. foreign aid programs. Agricultural scientists and research administrators have frequently traversed the nations of the Third World in an effort to convince the leaders of those countries of the wisdom of adopting the U.S. model for organizing agricultural research.

This diffusion effort has been based largely upon the personal experience of scientists and administrators and an idealized view of how the U.S. research system works. The organizational charts, the official statements of how teaching, research, and extension are interrelated, and the mechanisms for providing input from farmers into the research process often were taken as actual practice rather than as ideals toward which the research system should strive. Similarly, the goals, beneficiaries, and larger social context in which research was embedded remained taken for granted.

As a result, in many nations, the form rather than the substance of the U.S. agricultural research system was adopted. All too frequently, despite substantial investment in infrastructure and scientific training, research institutions have remained divorced from the people they were intended to serve.

In this volume we systematically examine the process and context in which public agricultural research is conducted in the United States. We also submit the very process of doing science to a critique that raises fundamental social, economic, political, and even philosophical issues for consideration and reflection by both agricultural scientists and laypeople. Our analysis does not substitute for experience, but it provides a broad overview that complements necessarily limited personal experience. It is our hope that such an analysis will increase our understanding of the processes and products of science, and of the relationships among science, agriculture, and the broader social, economic, and political forces of society. Furthermore, we hope that this knowledge will aid scientists, administrators, and policymakers in making the difficult decisions that lie ahead. Finally, it is our hope that such an overview will make it possible for scientists and administrators in developing countries to select more carefully those aspects of the U.S. model suited to their particular situation.

We are not, of course, the first to raise some of these issues and concerns. The publicly sponsored "Pound Report" (National Research Council, 1972), the recent report to Congress (U.S. Office of Technology Assessment, [OTA], 1981), and the privately supported work by Hightower (1973) have raised similar issues. However, our work differs on several fundamental points. First, we focus upon the world of the "bench" scientist, the social world in which the scientist operates. Second, we set the publicly funded research network into its social and historical context.

Our study draws upon a wide range of materials and methods. We conducted in-depth interviews with scientists at several institutions; analyzed mail questionnaires from over 1400 scientists; reviewed historical records; and perused official government and disciplinary association statements about research. Details on the precise methods used are provided for the interested reader in the methodological appendix. Of course, we make no claim to having exhausted the full range of strategies for understanding the research process. Studies of behavior in scientific settings, citation analyses of the scientific literature, and other approaches would doubtless yield different results. It is our hope that this volume will help to stimulate complementary research on the agricultural sciences.

We have written this volume with a broad audience in mind. We have tried to eliminate jargon and to explain technical terms when they seemed necessary. We have also avoided as much as possible statistical procedures with which educated laypersons would be unfamiliar. This proved to be both a difficult task and a learning experience for us. We found that terms with specialized meaning in our own field were understood differently in other fields, and that everyday words we used had technical meanings to others. This discovery served to underscore our concern with the disciplinary isolation that appears characteristic of much of contemporary agricultural scientific research.

In Chapter 1, we review the rise of public agricultural science in the latter part of the last century, focusing on certain key themes that continue to manifest themselves in contemporary research. The second chapter provides an overview of the contemporary agricultural research enterprise. It also sets the stage for the rest of the volume by elaborating upon the criteria scientists see themselves employing in their choice of research problems.

Chapter 3 provides a demographic profile of agricultural scientists, discussing in particular the influence of age, race, sex, family origins, and education upon research problem choice. In Chapter 4, we analyze the role of formal and informal communication in agricultural research; in Chapters 5 and 6, respectively, the disciplinary and organizational influences on research are examined. Extraorganizational influences (e.g. external funding and consulting) are discussed in Chapter 7. Chapters 8 and 9 address for whom and for what purpose research is conducted. They provide an account of answers given by scientists and compare these with official documents and issues raised by critics. Chapter 10 examines emerging issues for agricultural research in the decades ahead. Finally, in the concluding chapter, we discuss the implications of our findings for agricultural science policy. To date, there has been no consciously formulated and integrated agricultural science policy. Recent developments in Congress and elsewhere suggest that such a policy is likely to be framed in the near future. We conclude by examining the role of agricultural research in a democratic society.

The Rise of Agricultural Science

The contemporary structure of agricultural science in the United States, as in other Western countries, has been evolving for over a century. However, as Bonnen pointed out, "The thing that strikes one foremost out of the entire history of the evolution of these institutions is that at no one point, at the beginning or even well along in the process, did anyone have a completely detailed blueprint of what the system was to look like or what it was to do" (1962:1286). We begin by asking, How did the contemporary structure come to be?

One way of answering this question would be to detail the complex set of events leading to the development of the present system. Such a historical account could easily fill several volumes, and, to a large extent, it has been done by others (Knoblauch, Law, and Meyer, 1962; Rosenberg, 1976; True, 1937). Alternatively, our focus here is on those aspects of the history of agricultural science that are relevant specifically to the current process of formulating research problems.

The Socioeconomic Contexts of Agricultural Research

For thousands of years, farmers have tilled the soil and husbanded animals. They domesticated animals, invented various mechanical devices, and selected high-yielding plant varieties. They built irrigation networks and developed methods of pest control. Then, a little more than one hundred years ago, agriculture became a subject of scientific investigation. The entry of science into agriculture brought with it enormous increases in productivity, the sharing by the entire world of the immense pool of plant and animal resources (Spitz, 1975; Brockway, 1979), and the possibility of an industrial and even post-industrial society. Why did this extraordinary shift occur not only in the United States but in the world as a whole? It is to this question that we turn first.

The eighteenth and nineteenth centuries were a period of rapid expansion by the major world powers. Britain and France divided much of the old world into a patchwork of colonies, while the United States extended its boundaries over North America and its hegemony over Latin America. The ideology of the "white man's burden" prevailed, but there is little dispute today that this ideology was designed in part to conceal the economic exploitation of large areas of the world.

Not surprisingly, a major feature of this expansion was agricultural. Everywhere one might have traveled in the colonies, agriculture predominated; manufacturing was reserved for the "mother countries." Yet, this was a new kind of agriculture quite different from the peasant agriculture of Europe. Farming was changed fundamentally from a means of subsistence to a commercial enterprise. Unlike the peasants who often lived marginally and were concerned with the *maintenance* of a small plot of land for their own subsistence and the payment of taxes, the commercial farmer or plantation owner was concerned with rapidly increasing production. Moreover, the owner was prepared, financially and spiritually, to take the necessary risks to expand production. Indeed, so concerned were the new farmers with increased production that by the early nineteenth century large areas of the southeastern United States had already been abandoned. In short, at the same time as it freed the farmer from the limited annual return from subsistence farming, the entrepreneurial capitalist mode of production also tended to break the ties between the farmer and the land. Land was transformed into a resource to be used until depleted and then abandoned. As a result, the worldwide volume of agricultural commodities grown for off-farm use and for export increased steadily. Until the late nineteenth century, however, much of the increase was attained by increasing the land area devoted to the production of marketable crops.

It was far more profitable to increase the area under cultivation than to use the land more efficiently through improved cultural practices. Furthermore, the dispersed ownership structure in agriculture made private investment in research by farmers unprofitable. Therefore, there was little economic incentive for agricultural research other than that on machinery. However, wealthy farmers soon discovered that by making research the responsibility of the state, they could reap short-run benefits and avoid most of the costs.

Among the most vocal proponents of state support for agricultural research were those who were most likely to benefit from research. As economist Vernon Ruttan explained:

Under competitive market conditions the early adopters of the new technology in the agricultural sector tend to gain while the late adopters are forced by the product market "treadmill" to adopt the new technology in order to avoid even greater losses than if they retained the old technology. One effect of the treadmill phenomenon is . . . to limit the economic motivation for support of agricultural research to a relatively small population of early adopters of new technology. The early adopters also tend to be the most influential and politically articulate farmers (1980:540).

In short, through agricultural research, profits of early adopters could be increased at the expense of late adopters and profits could be increased within a region at the expense of other regions. At the same time, the costs of developing innovations could often be passed on to the state. Once early adopters or one region successfully implemented an innovation, prices for farmers' products dropped. Late adopters and growers in other areas had to innovate merely to hold on to their share of the market.

For the U.S. farmer, the market was large. Cotton and tobacco had been exported to Europe since colonial days, and the development of railroads to the midwest opened the possibility of adding both grain and livestock to the exports. As G. E. Morrow, acting president of the Association of American Agricultural Colleges, put it, "In a greater degree than ever before the American farmer is feeling the effects of direct competition with a vastly increased number of fellows in his own land and of many millions in many other lands" (1895:26). Thus, very rapidly, the possibilities for worldwide commercial, as opposed to local subsistence, farming were made available to many U.S. farmers. Nevertheless, the idea of government-sponsored research was foreign to the thoughts of most Americans in the mid-nineteenth century.

The Origins of Agricultural Science and the Baconian Ideal

Perhaps the most fundamental issue to be resolved before there could even be agricultural sciences in the United States was the issue of government involvement in agriculture. Until the mid-nineteenth century, science was pursued in the Western world in a Cartesian style. The image of the scientist was that of the isolated, hard-working individual who composed grand and elegant theories and only later, with the help of a staff of technicians and apprentices, tested them by observation and experiment. Through careful attention to method, the scientist could discover the laws (of God) by which the universe was ordered. The scientist's role was much like that of the atomistic laissez-faire entrepreneur (Haberer, 1969:72). He worked in competition with other scientists—often in secret in furtherance of his own self-interest—but, conveniently, the community as a whole benefited from the discovery of new truths.

The Baconian model, on the other hand, was entirely corporate. The members of the House of Salomon (the scientific community in Bacon's utopian novel *New Atlantis*) worked together as a group supported by the state to further the material welfare of the population. They required no particular genius to construct an elaborate system. Instead, they placed their trust in the power of the scientific method to overcome their subjectivity. The results they produced were utilitarian, immediately applicable to the problems of the state and the larger society. This corporate scientific community permitted the growing domination of nature (Leiss, 1972).

The suggestion that agricultural science should be institutionalized shifted the balance in favor of Bacon's model of science in three ways: First, the concept of an agricultural science was explicitly utilitarian; knowledge was to be pursued for its usefulness in improving the material conditions of the population. Second, it was to be pursued by large numbers of persons who had learned the proper methods. Genius was helpful, but not necessary. In fact, less than 40 years after the founding of the agricultural colleges, there were nearly 22 times as many agricultural scientists as physicists in the United States (U.S. Office of Education, 1901; Forman, Heilbron, and Weart, 1975). Third, it emphasized empirical results over theory. As Congressman Hatch put it, "A scientific experiment is made, not for the purpose of seeking or sustaining a theory, but of learning a fact" (1886:4). While today some may still debate the degree to which government should be involved in agriculture, the question of whether it has any legitimate role to play appears to have a self-evident answer. The House of Salomon is an accepted part of U.S. society.

This was certainly not the case in 1835, when Henry Ellsworth was appointed Commissioner of Patents under President Andrew Jackson. Ellsworth urged increased involvement of government in agriculture and received in 1839 the "grand" sum of $1,000 for collecting agricultural statistics. The Southern farm journals became quite alarmed at what they considered to be inaccurate data collection. More importantly, Southern journalists and legislators felt Ellsworth's project to be an infringement of states' rights and clearly unconstitutional. In 1845, for reasons still unclear but probably related to his bold advocacy of government support, Ellsworth was forced to resign.

Through the 1850s, a similar controversy focused on Senator Morrill's proposal to establish land grant colleges for "agriculture and the mechanic arts." Wealthy farmers and editors of many agricultural journals—not scientists, who remained aloof—were convinced that farming based on von Liebig's agricultural chemistry would be far more productive than traditional farming. Again, opposition from the South on constitutional grounds prohibited passage of the bill. Then negotiation over another issue, slavery, broke down, and the South seceded. Morrill and his associates found that their opposition had disappeared. In 1862, the Land Grant College Act was passed. Each state was provided with a grant of federal land, proceeds from the sale of which would provide an endowment for an "agricultural and mechanical" college.

Among the host of new issues raised by passage of the Morrill Act, a primary question was whether the agricultural colleges should do research at all. Until the 1860s, it was universally assumed that what a professor did was teach (Rosenberg, 1971). What little research had been performed in agricultural science before the Morrill Act was of virtually no consequence. As Regent J. M. Gregory of Illinois Industrial University put it in 1869: "Looking at the crude and disjointed facts which agricultural writers give us, we come to the conclusion that we have no *science of agriculture*. Botany is a science—chemistry is a science—but agriculture is not a science in any sense. . . . It is simply a mass of empiricism" (Convention of Friends, [1871] 1967:ix). Faculty at most colleges quickly realized that research was necessary if the colleges were to prove their worth.

The questions, What kind of research to do? and What constitutes research? were far more difficult ones to answer. Substantial insight into these problems can be gained by examining the proceedings of the Convention of Friends of Agricultural Education ([1871] 1967). This organization, a forerunner of the National Association of State Universities and Land Grant Colleges (NASULGC), spent several days in debate over just what constituted a proper experiment. What size should the fields be? How could one generalize an experiment made in Maine to one in Mississippi? Was field research necessary at all, or could most relevant questions be resolved in a laboratory? And, on a more esoteric level, was agriculture a science or an art? Of course, these debates were not resolved at the Convention.

A related, and still unresolved, debate opposed the advocates of basic and applied research. The faculty of the new colleges consisted largely of persons with education in the basic sciences, although a number of good farmers were also hired. These early agricultural scientists desired to acquire a status equal to that of physicists, biologists, and chemists; but if they had not been able to demonstrate the immediate utility of their research, their sources of financial support would soon have dried up. Furthermore, while college faculty agreed on the need for more research funds, they disagreed among themselves about what constituted research. The newly founded journal *Science* editorialized that scientific work was necessary before any practical advances were to be made in agriculture (*Science*, 1884). E. W. Hilgard, a noted soil scientist, responded in a letter that science and practical work should not and could not be separated (1885). The editors replied that "there appears to us to be comparatively little danger that the work of American experiment stations will be too rigidly scientific and too far removed from the apprehension of farmers" (*Science*, 1885:21). Indeed, the split between those who advocated "original research" and those who desired "practical information" became a central issue in the formulation of the Hatch Act of 1887, which established the State Agricultural Experiment Stations (SAES) (Knoblauch, Law, and Meyer, 1962). Although the wording was finally made vague enough to please everyone, the SAES were decentralized. Each was responsible to its respective state legislature, which, in turn, was eager to show results to its constituents. This assured the "victory" of applied interests.

If researchers had little idea of what agricultural research should be, their constituencies had even less. To the public, research was virtually indistinguishable from regulation and inspection. Chemical fertilizers, then proliferating much like patent medicines and with equally fantastic claims, were frequently brought to experiment station scientists for evaluation (Rosenberg, 1971). On the other hand, geneticists and nutritionists had difficulty in securing funds for rats, mice, and guinea pigs. Critics were appalled at their requests and asked why state funds should be used to breed pests (Rosenberg, 1971). Scientists were expected to answer all questions, respond to a deluge of letters—one director reported receiving 75,000 letters per year (R. Scott, 1970)—give speeches, attend farmers' meetings, and otherwise be available when anyone had a problem. This was a situation that Bacon never envisioned.

To better educated and more scientifically oriented station directors and faculty, the situation was, at best, barely tolerable. "The able and articulate among experiment station scientists and administrators had obviously to respond by seeking to restructure an environment which seemed on every hand to compromise their autonomy. Idealism as well as self-interest at once prompted and legitimated such policies" (Rosenberg, 1971:6). In an attempt to secure additional funds and to escape from the paperwork and testing that substituted for research, administrators became research entrepreneurs. The strategy they employed was fairly straightforward: "the creation of institutional strength through alliance with members of the business, agricultural, and political communities" (Rosenberg, 1971:12–13). One example was the intimate relation between Harry L. Russell (Dean of the Wisconsin College of Agriculture from 1907 to 1930) and the banking community: "Russell possessed a considerable amount of business experience and many business friends. It seemed natural to him, as dean, to appeal to those associates to help him advance agriculture in Wisconsin" (Beardsley, 1969:65). Indeed, Russell sought and obtained direct financial support for a variety of projects from the banking community. At the same time, he attempted to avoid direct interference by those groups in the affairs of the college (Beardsley, 1969). Like many of his contemporaries, he viewed small, local banks and businesses as unconcerned with the internal affairs of the college but committed to the prosperity of their local community. On the other hand, he feared the plans of the Rockefeller-backed General Education Board and Sears Roebuck for privately financed extension services that would exploit the farmer for "advertising purposes" (Russell, 1912).

Despite these difficulties, the late nineteenth and early twentieth centuries were a period of increasing professionalization for the agricultural sciences. By the late nineteenth century both agricultural scientists educated in land-grant agricultural colleges and others trained in Germany began to enter the SAES. Through the help of A. C. True of the Office of Experiment Stations (OES), the influence of experiment stations and colleges on their respective state representatives, and the emerging influence of the scientific disciplines themselves, the Adams Act, which provided funds to be used only for "original research," was passed in 1906 (Rosenberg, 1964). The act's provisions for project approval within the federal Office of Experiment Stations made it far more difficult to divert funds for other uses. (Experiment station budgets were rather substantial, and impoverished state colleges frequently dipped into them for teachers' salaries and even insurance premiums [Rosenberg, 1971].) Moreover, a project system originated by True made it possible for the Department of Agriculture to veto inappropriate uses of funds.

Nevertheless, experiment station scientists still found themselves bogged down in a bewildering variety of what would later be called extension activities (Storer, 1961:46–47). The position of researchers within the experiment stations about 1900 is graphically depicted in Figure 1.1. At a time when researchers still sought to define their respective research "territories," many often conflicting demands were

FIGURE I.I
Sources of Influence on Agricultural Research in the State Agricultural
Experiment Stations ca. 1900.

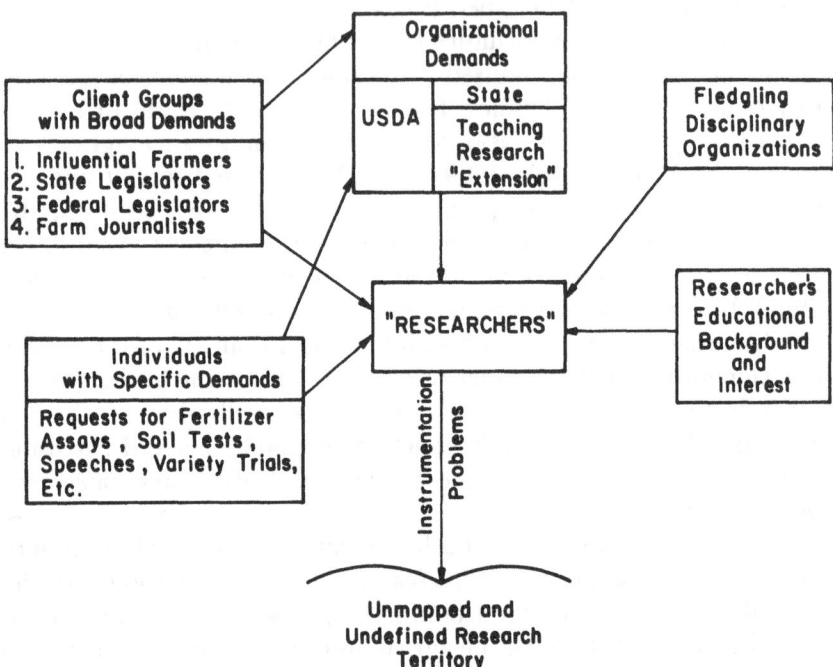

placed upon them. Moreover, no formally defined channels were available either for making requests of researchers or for disseminating research results. A bewildering array of bulletins, reports, pamphlets, and leaflets were developed and mailed to farmers. Researchers went on tours, often spending weeks at a time traveling through the countryside. In short, there were not, as yet, any defined research priorities or dissemination pathways.

Agricultural Science Comes of Age

If the nineteenth century can be characterized as a period in which the agricultural sciences were developed in response to pressures from various interest groups, then the first two decades of the twentieth century must be seen as a period in which agricultural scientists began to shape and reshape the very social environment that had created their fields of endeavor. By 1900, despite myriad problems, agricultural science was firmly rooted. The deans of U.S. colleges of agriculture, many of them now land-grant college graduates themselves, were politically important men who increasingly saw themselves allied with scientists and engineers in other fields as part of the vanguard of a movement that would revolutionize not only agriculture but business, industry, and even politics.

The new movement wanted to do more than merely increase productivity. Although concerns about productivity and the level of farm exports were to be considered, it was felt that these issues could be resolved only by a thorough reorganization of rural life. As Kenyon Butterfield of Massachusetts Agricultural College put it, "There must be a union of the country school teacher, of the agricultural college professor, of the rural pastor, of the country editor, with the farmers themselves for the production of an increased crop of *new farmers*" (1907:65, emphasis added). The program for change was perhaps best set forth in the report of the Commission on Country Life ([1909] 1911). That commission, chaired by horticulturist and Cornell University Dean Liberty Hyde Bailey and made up largely of leading agricultural scientists of the day, summed up the scientists' program to create "a highly organized rural society." First, sociologists and economists would be hired to conduct surveys of agricultural conditions around the country. Social scientists would provide "the facts" upon which rural reorganization would be based. One "rural organization specialist" suggested that "the rural survey may be compared to the inventory taken periodically by the modern merchant" (Thompson, 1917:129). Then, through a national extension service with agents in each county, requests for information would be organized into a coherent, national system. The farmer would be provided with the answers obtained through science. This was important to the research process, to relieve scientists of the barrage of correspondence from farmers. Finally, with the reorganization and consolidation of the rural schools and churches, the improvement of rural highways, the adoption of sound business practices by the farmer, and the development of rural leadership, rural life would be transformed.

Concern with directed social change was, of course, not unique to agricultural scientists but dominated the entire social world at the time. The Taylorists' desire to introduce scientific management into the workplace, home, and school (Taylor, 1911), the home economists' enthusiasm for scientific management as the end to household drudgery (Jones, 1917), the conservationists' interest in insuring maximum sustainable yields (Hays, 1959), and the Progressives' drive to reform city government and the schools (Spring, 1972) and to develop social welfare legislation (Piven, Fox, and Cloward, 1971)—all shared certain themes with the agricultural scientists' plans to reshape society. C. R.

Jones, dean at West Virginia, clearly perceived the link between Taylorism (scientific management) and agricultural science:

> The farm management movement does not seem to have been directly influenced by the Taylor system. Indirectly it has greatly advanced the movement by its effects on the plants which manufacture farm machinery and farm implements, and probably through the stimulation which the Taylor system gave to cost accounting and efficiency of production. This influence has been unconscious rather than conscious. Yet the movement has had a logical evolution and the trend is toward the same ideal as that advanced in the Taylor system (1917:111).

The movement to establish "engineering experiment stations" in conjunction with those devoted to agriculture developed simultaneously with Taylorism. First in 1897, again in 1907, and for a third time in 1916, abortive attempts were made to secure federal legislation to this end (Potter, 1923). A section within the Association of Agricultural Colleges was established to deal with engineering. Several states even established engineering stations with state funds. And all appear to have embraced essentially the same aims as the Taylorists.

The key ingredient for scientific management was organization. The older laissez-faire economy that had served during the nineteenth century was to be replaced. Eugene Davenport of the University of Illinois expressed this attitude to his fellow station directors and college presidents: "We have now reached a time in world development when we recognize the fact that many good things cannot take care of themselves but must be cared for, and that the policy of laissez-faire is often fatal to peaceful progress" (1919:12).

The replacement for this now "archaic" system was to be a corporate one in which large-scale organizations would plan for the production and distribution of goods and services. The trusts created in the late nineteenth and early twentieth centuries were seen as a model for the new order, the last step in social evolution (Hooks and Flinn, 1981). Indeed, Davenport carried his point to an uncomfortable extreme, suggesting that

> many individuals will be crowded out as agriculture exacts more knowledge and skill. . . . The great laws of evolution and the survival of the fittest will continue to operate, and, in the interest of progress, they ought to operate. Progress is not in the interest of the individual, and it cannot stop because of individuals. Everything must surrender to the central idea that this is a movement for the highest attainable agriculture in the fullest possible sense of the term (quoted in Danbom, 1979:40).

Although not all of his colleagues would have agreed with him, for they valued the family farmer as the kingpin in the maintenance of democracy, many shared his firm distrust of the "disorganized" countryside and fear of class warfare. They needed only to look to Eugene V. Debs' rising popularity as a Socialist Party presidential candidate, to the myriad and often violent strikes by industrial workers, or to the volatile scene in Europe to confirm their worst fears. As C. A. Duniway, president of the University of Wyoming, put it, "There is a real danger that class consciousness and class interests may produce chronic struggles for supremacy" (1915:96).

In their zeal for science, researchers sought to apply it not just to technical matters but to the resolution of all conflicts. Confusing objectivity with political consensus, they saw science as the supreme mediator among competing classes and interests. In a paper on "The Authority of Science," W. H. Jordan, director of the New York Experiment Station, noted: "The serious side of the world-wide movement is the conviction that science is a trustworthy guide in directing our activities. In all ages man has been prone to seek the guidance of authority. . . . but now he has turned to science, and, excepting in things spiritual, it utters the final word" (1907:62). Similarly, Bailey could argue that "only as we found government on evidence and develop it in the scientific spirit can we expect to have really good government or to make the best progress in civilization" (1910:27). Science might not be infallible, but it was far superior to politics and class conflict. Scientists' views were supported by Theodore Roosevelt, whose "administration and his social and political views are significant primarily for their attempt to supplant [social] conflict with a 'scientific approach' to social and economic questions" (Hays, 1959:267).

Bailey and Gifford Pinchot of the Forest Service were, perhaps, the central figures in this movement. As chairman of the Country Life Commission, member of the American Association of Agricultural Colleges and Experiment Stations, and president of the American Association for Agricultural Legislation, Bailey was well known and influential. His many books and edited collections were read throughout the nation. Bailey promoted the application of science to the political sphere. In at least one instance, he made it quite clear that society could be based upon either conflict or consensus and strongly argued for the consensus position (Bailey, 1920; see also Davenport, 1919; Wiest, 1923). In his view, this consensus was not to be achieved by discussion or by democratic procedures: "It should be deliberate. We should know before we legislate. There should be a given body of men and women, designated in an organization, to which one may go for facts and for advice founded on facts rather than on sky-blue opinion" (Bailey, 1920:117–118). Similar claims were voiced by Pinchot: If timber companies, farmers, and ranchers argued over the appropriate uses of forest lands, the Forest Service would, through the application of science, mediate those claims (Hays, 1959:58).

To summarize, although the origins of the movement are not entirely clear and must await further historical research, it appears to have arisen, in part, from: (1) a desire on the part of scientific leaders to avoid conflict, especially class conflict; (2) a concern for the rapid depletion of resources occasioned, in part, by the close of the frontier; (3) a realization that the United States was now part of a complex international economic order in which it needed to continue to compete effectively; and (4) an increasing skepticism about the ability of the laissez-faire economy to produce the good society. As a result of these concerns, many agricultural scientists felt compelled to attempt a transformation of U.S. society.

There was, of course, no malice in this desire to change society. At the time, it appeared to everyone that science and organization would make a better world for all: "The new farmer has his largest conquests yet to make. But he has put his faith in the strong arm of science; he has at his hand the commercial mechanism of a world of business. He believes he will win because he is in league with the ongoing forces of our civilization" (Butterfield, 1907:54). But, alas, the world is never so simple or automatic. There were serious flaws in this reasoning. Scientists and research administrators "seldom considered that a thoroughly scientific and mechanized agriculture producing standardized products might well make farming much more of a business than a way of life, drive many people out of it, and bring to agriculture the problems of class differentiation and conflict that plagued industry" (Danbom, 1979:66). Although they might be frustrated by the farmers' often stubborn adherence to tradition, the agricultural scientists of the turn of the century nevertheless saw those very farmers, and the relative independence that so clearly distinguished them from the peasants of Europe, as the most important element of a desirable society. That this very element would be destroyed by increases in productivity and efficiency was, as yet, little understood.

A New Kind of Science

With the close of World War I, farmers suddenly found themselves with huge surpluses and glutted markets. Throughout the 1920s, while the urban population enjoyed enormous prosperity, the "farm problem" continued. It became apparent to station directors and college deans that the problems faced by farmers were at least as much social and economic as they were technical, but the experts disagreed over the nature of the problems as well as the corresponding solutions. Many argued that the "farm problem" was a matter of class structure. As rural sociologist W. F. Kumlein presented it to the station directors, society was divided into three great classes—capital, labor, and agriculture:

A study of the organization and mode of procedure of these three economic classes reveals some interesting comparisons. In the early history of our country, individualism was the order of the day in each of the three classes. Capital was the first to modify individual competition within its own group and to see the advantage of capitalistic combinations. Out of this change grew large-scale management, which resulted in eliminating competition, regulating output, and stabilizing prices.

Labor in self defense came to see the weakness of individual bargaining. Gradually labor, through organization into unions, evolved the policy of the closed shop, restricting its output and to a remarkable degree succeeding in stabilizing its wages in keeping with a fair standard of living.

While considerable progress has been made in the last quarter of a century, agriculture has been the least successful of the three classes in developing collective bargaining power (1927:150).

This argument seemed so reasonable to his colleagues that John Lee Coulter, president of North Dakota Agricultural College, felt compelled to point out that organization alone was insufficient to resolve the farm problem (1928).

Thomas Poe Cooper, director of the Kentucky Experiment Station and chairman of a "Special Committee on the Agricultural Situation" suggested another cause of the farm problem. "A very important factor, which is retarding the recovery of agriculture, is found in high marketing, processing, and distributing costs of farm products which have persisted since the deflation period [after the First World War] and which, in some instances, have increased since that time" (Cooper et al., 1927:93).

Other scientific leaders, however, saw the roots of the problem from an entirely different perspective. Albert F. Woods, president of the University of Maryland, in his presidential address to the Association of Land-Grant Colleges in 1925, blamed the farmers themselves: "Non-paying 'boarders' of all kinds—crops, machinery, acres, and livestock—must be eliminated. The low-grade inefficient farmer who has demonstrated inability to learn and cooperate with others must be eliminated. These produce the surplus by slovenly methods and do most of the howling" (1926:27). Nor did Woods confine his criticism to individual farmers,

> for a community made up of such [inefficient] individuals repels the efficient individuals. The inefficient individuals gravitate to it. Low efficiency results in low income; low income means low taxable basis; this means poor schools, poor roads, poor police and health protection— in short, degeneration. Nothing good comes out of such places; nothing good goes in. They are social and economic liabilities rather than assets. As they increase in number, the state is weakened.
>
> On the other hand, a good community attracts good individuals. They seek to improve their individual efficiency and cooperate with each other to accomplish what they cannot do individually. They know the value of education, and organize and support good schools. Their greater intelligence results in increasing efficiency and larger income. They invest it intelligently to improve their business, their communities and their homes. Such communities are assets of increasing value to the state (1926:28).

While Woods' position may well have been abhorrent to some of his contemporaries, it was supported by others. In the following year, agricultural engineer J. Brownlee Davidson argued: "At the present time, about 25 percent of the population is engaged in agriculture and this number could be materially reduced. . . . This should not be looked upon as a national calamity. It would simply be a continuation of the history of the past century. It should be looked upon as progress" (1927:182). Similarly, at a 1927 meeting of delegates of the American Society of Agricultural Engineers (ASAE) with President Coolidge in Rapid City, North Dakota, eight recommendations for resolving the farm problem were made, including "Surplus population should be transferred to industrial activities" (Stewart, 1979:32–33). Robert E. Stewart, ASAE's official historian, notes: "These recommendations made good sense, although the one related to 'surplus people' has a callous sound to our squeamish present-day ears, a totalitarian flavor which Zimmerman and friends probably never intended. Engineers sometimes use inhuman phrases when proposing solutions to human problems" (1979:83). In short, the victims were to be blamed as the perpetrators of low prices; and what had been looked upon as a worrisome trend was to be regarded as another manifestation of progress.

Similar disagreement surrounded strategies for resolving the "farm problem." Andrew Boss, a farm management specialist from Minnesota, while convinced that big business was inappropriate to agriculture, encouraged increased specialization: "Successful commercial production demands specialization and concentration" (1927:145). The Committee on the Agricultural Situation concurred that "the application of science and technical skill to agricultural production favors specialization . . ." (Cooper et al., 1927:107).

E. D. Merrill (1928), dean at the University of California, favored organization. W. F. Kumlein was even more forceful in asserting that "agriculture has reached the place where it has no alternative left but to organize itself to compete on a similar basis with capital and labor or else be exploited as a class" (1927:151).

Still others favored massive government intervention: "It seems to me self-evident that government policies have in the past, and will in the future continue to have, a profound influence upon the development of a permanent and prosperous agriculture" (Mumford, 1926:192). From this perspective, intervening in the marketplace, reducing production, and planning by the national government would resolve the farm problem.

Yet, even as their solutions differed, the views of these agricultural science leaders converged in two central ways. Nearly all were agreed that the problem was social and economic in nature; they also concurred that the development of a successful strategy for resolving it demanded the application of science—economic science. As a result, they began to hire more economists. By 1930, the number of social scientists (most of whom were economists) employed in the experiment stations and USDA had risen to 823 (Schultz, 1941:175). At the conclusion of the following decade, 1,461 social scientists were employed in those same institutions (Schultz, 1941:176).

The Increasing Role of Social Scientists

Nearly all of the early agricultural scientific leaders received their training in the natural sciences. A few, such as Bailey and Butterfield, dabbled in economics and sociology, but rare was the scientist with formal training in these fields. The new directions formulated during the first two decades of this century, however, demanded a better knowledge of the economy and society; planning required a new kind of scientific knowledge.

Most of the early agricultural economists and rural sociologists were men recruited from other agricultural disciplines. Economist Milburn Wilson and sociologist Dwight Sanderson, typical of pioneer agricultural • social scientists, were both trained in the natural sciences.

The Department of Agriculture had separated economic from technical work as early as 1913, but most of the early economic reports were devoid of economic theory and essentially statistical in character. With the passage of the Purnell Act in 1925, however, departments of agricultural economics and, less commonly, rural sociology were formally established at many land-grant schools. These new departments espoused a kind of social science quite different from that of their counterparts in the liberal arts. Although many agricultural economists have assumed that their discipline consisted largely of economics applied to agriculture (Yang-Boo, 1978), these early scientists were characterized by a penchant toward empiricism and an unfamiliarity with and even lack of interest in the bodies of theory developed in Europe. Even as late as 1941, Theodore Schultz could write with apparent approval, "The experiment station has provided for them [rural sociologists and agricultural economists] a research environment and facilities for adding new facts to the coral reef of knowledge" (1941:3). Not only did these agricultural scientists differ substantively from their counterparts in the mainstream, but they were formally separated from them. Agricultural economics began as a discipline separate from economics; after some years of bickering, rural sociologists broke away from the American Sociological Society and formed their own professional association (Lacy and Busch, 1982a).

The emphasis by social scientists on "fact gathering" was encouraged in several ways. Within the land-grant colleges, "they were under heavy pressure to demonstrate the utility of their offering, since they were judged by the same concept of utility held by the dominant physical scientists of the experiment stations" (Brunner, 1957:5). In short, those natural scientists and administrators who encouraged the development of the social sciences saw their role as essentially the extension of the scientific method to the social and economic spheres. Moreover, social scientists themselves were often of the opinion that the social sciences could become truly scientific only if the methods of the natural sciences were employed. Therefore, many eagerly embraced the reductionism and empiricism typical of the natural sciences at the time.

The economists quickly took up the call that had been voiced earlier by station directors and college deans, though they expressed it in somewhat different form:

> As scientists, these men had roots in the growth of the new universities, the development of scientific agriculture, and the conservation movement as well as in the rise of the social sciences. Their work, however, was also influenced by nonscientific traditions, especially the traditions of American business and American democracy. These men attempted to make the farm business profitable; they even encouraged farmers to imitate the practices of urban businessmen in hopes of accomplishing

this goal. Rejecting elitist theories, they promoted the growth of farmer committee systems designed to encourage popular participation in planning and administering farm policy (Kirkendall, 1966:256).

Although they were by no means of one mind, these social scientists and economists assumed as their own the program epitomized by the report of the Commission on Country Life. At the same time as their natural science colleagues were becoming more specialized, they maintained a concern for issues beyond productivity—issues of social welfare, planning, and democracy (Finegold, 1982). As a result, they became both more visible and more subject to political pressure.

Those who were most outspoken on these issues during the early 1930s were eventually forced to leave because of the pressures brought to bear upon them (Hardin, 1946). Rexford G. Tugwell resigned from USDA after his stance made him a political liability to Roosevelt's reelection in 1936 (Kirkendall, 1966). A decade later, H. R. Tolley resigned from the Bureau of Agricultural Economics (BAE). As a result of intense pressure from the Farm Bureau, he felt the BAE could no longer "function as an independent scientific institution" (Kirkendall, 1966:241). Similarly, anthropologist Walter Goldschmidt found himself the center of controversy even before he began a study that was later to document the adverse effects of large-scale farms on community structure. As a result of pressure exerted by the Farm Bureau, the USDA refused to publish his report. Looking back on the incident, Goldschmidt bitterly concluded, "thus agribusiness interests have successfully curtailed those research activities not devoted to economic profitability and the techniques of production and marketing" (1978:486).

The problems involved in engaging in "action programs" were publicly discussed at the annual meeting of the Association of Land-Grant Colleges and Universities in 1938. R. E. Buchanan, director of the Iowa Agricultural Experiment Station, expounded the quietist position: "Certain of the institutions constituting the membership of this association believe that they lose one of their most precious endowments, academic freedom, including research and educational freedom, if they 'implement' any direct action agricultural program. These schools are convinced that undesirable political domination is a sequel . . ." (1938:113). He concluded by asserting that executive functions should be avoided and that the universities should stick to education and research in the more narrowly defined sense. Five years later, Edmund E. Day, in a presidential address to the Association, put it even more bluntly: "The function of these institutions is to accumulate and disseminate knowledge, not to supervise, or regulate, or reform practical operations" (1943:25).

From the mid-1930s on, most agricultural social scientists employed by USDA and the land-grant universities moved away from their earlier activist stance. Theodore Schultz, then head of the Department of Economics and Sociology at Iowa State, could sadly conclude in 1941 that "certainly there is a strong disposition to avoid the more important issues if they are controversial and the most significant problems in the social sciences are of this type" (1941:170). Ironically, Schultz himself resigned several years later as a result of pressure from local dairy interests (Hardin, 1955).

While agricultural economists avoided policy questions by focusing on forecasts, farm management, and marketing, rural sociologists tended more and more to avoid agricultural issues altogether. The voluminous literature on rural communities often ignored the agricultural base upon which those communities were founded. Other sociologists found safe havens in demography and studies of rural health. Still others enthusiastically embraced the products of the natural sciences and attempted to devise strategies for their diffusion (Rogers and Shoemaker, 1971; cf. Busch, 1978; Lacy and Busch, 1982a).

In short, with few exceptions, "programmatic research" (Hardin, 1947) has been avoided since the late 1930s. There is no journal of agricultural policy, and no policy research unit within USDA. Similarly, the National Association of State Universities and Land-Grant Colleges has tended to avoid policy issues. Writing in 1955, Charles Hardin noted that "the committees of the association never returned to the critical self-appraisal of the orientation of experiment station research that characterized the 1920s" (1955:153). Indeed, the very role of agriculture in the association diminished over the years. By 1960, the experiment stations had only one vote in 90 in the association senate (Volk, 1960). In the early 1960s, the opening of the association to non-land-grant institutions, most of which had little or no interest in agriculture, further reduced its effectiveness as a forum for issues of agricultural policy. Today, representatives of the Division of Agriculture of NASULGC have only four of about 160 votes in the association senate (Hudgins, 1981).

The Construction and Reconstruction of Political Support

In addition to the internal forces shaping the structure of the agricultural sciences, negotiations also occurred with a wide variety of organized interests. Among these were the United States Department of Agriculture, whose role in agricultural research has changed greatly over the last century, and a variety of farm and business groups, some of which have had unique relationships with the agricultural colleges. First, let us briefly examine the role of the U.S. Department of Agriculture.

U.S. Department of Agriculture

When the agricultural colleges were first established, many assumed that the agriculture department would function much like the post office and war departments. Its commissioner (later secretary) would preside over a centralized agency that would respond to particularly urgent research needs of farmers. Indeed, in the quarter century following the passage of the Morrill Act in 1862, the Department of Agriculture tried several times to carry out programs to eradicate certain diseases and pests. All these attempts failed miserably.

When the Hatch Act was passed in 1887, it "changed the Department of Agriculture from a single central agency into a nexus of a system of semi-autonomous research institutions permanently established in every state" (Dupree, 1957:170). Each station was provided $15,000, unconditionally, for the purpose of conducting agricultural research. As a result, the SAES were autonomous from USDA. Not surprisingly, with farmers demanding immediate practical results and state legislatures wary of basic research, the Department of Agriculture's input was limited. The Office of Experiment Stations (OES) within USDA took on the limited roles of (1) providing through its publications a clearinghouse for research results, (2) attempting to place high quality agricultural scientists within each of the respective state stations (Rosenberg, 1971), and (3) using its audit powers to encourage "abstract" research and to discourage the mere collection of data.

In 1955, the formula for allotting Hatch funds was modified. Thereafter, a portion of the allotments was based upon the ratio of a state's rural and farm population to the national total and a matching contribution from the states was required. In addition, the research capabilities of USDA have expanded over the years. Nevertheless, the relationship between USDA and the SAES remains much the same today as it was in 1887.

Although an examination of the role of extension is beyond the scope of this volume, it is worthy of note that the establishment of the extension service in 1914 marked an increase in USDA control if only within that arena. Today, extension agents are federal civil service employees responsible, *in principle*, to Washington (see Lacy, Pigg, and Busch, 1980). On the other hand, teachers and researchers are state employees, and research money may originate in either state or federal treasuries. As a result, in each of the agricultural colleges, the three functions of teaching, research, and extension are separately administered; and a different set of power relationships marks each of the three activities. The fact that portions of a single individual's salary may be paid from three separate funds further complicates the issue. Popular writers and critics who speak of *the* relationship between the agricultural colleges and the USDA gloss over what is a complex, multifaceted set of relationships.

Farmers' Organizations

The relationship between the agricultural colleges and other constituencies is not as complex, but it is, nevertheless, unique. It is not at all uncommon for organized interest groups to lobby successfully for the formation of a special government agency to serve their needs. It is unusual, however, for a governmental agency to create a constituent organization. Yet, this has happened at least twice in the history of the American agricultural colleges.

In 1866, with the Civil War over and the South in ruins, southern agriculture was in extremely poor condition. President Johnson consulted with the Commissioner of Agriculture, Isaac Newton, who sent a clerk, Oliver H. Kelley, to the South to study agricultural conditions. After completing a tour, Kelley, a Mason, returned and suggested the creation of a secret, ritualistic farm organization for educational and social purposes. Newton liked the idea, and the Grange was created. Kelley was sent on tour throughout the country organizing local Grange chapters. Although the Grange proved ineffective as an educational organization, it did provide a new, organized constituency for agricultural education and research (True, 1929).

By 1890, Grange membership had dwindled, reducing the effectiveness of the organization. The Populist movement of the 1890s, which partially took its place, posed a threat to the banking and business communities. In an effort to reduce the rural discontent voiced by the Populists, the Rockefeller-endowed General Education Board sponsored the pioneering work of Seaman Knapp in farm demonstrations.

Knapp headed the Farmers Cooperative Demonstration Work division of USDA, the forerunner of the Extension Service. The American Bankers' Association appointed a committee on agriculture in 1909. And, as we noted earlier, the Country Life Commission pushed for national extension work. As McConnell puts it, "The [bankers'] program for agriculture included good roads, soil fertility, and education. Education meant cooperative demonstrations and county agents. This was the positive alternative to dissent and radicalism which the bankers held out to the sometimes misguided man of the soil" (1953:31).

The American Association of Agricultural Colleges and Experiment Stations, strongly supported by research faculty, saw in extension the opportunity to expand greatly the scope of the colleges while freeing researchers from the time-consuming process of dealing directly with farmers. Although the colleges desired both the extension funds and complete autonomy for their use, they had to compromise. Under the terms of the Smith-Lever Act (1914), agents' salaries were paid in part by federal, state, and county governments as well as, in some cases, by private organizations. How the agents were to divide their loyalties among four different organizations was never quite clear, and remains unclear even today (Lacy, Pigg, and Busch, 1980). What was clear was that county agents were instructed to work with groups and quickly became the organizers of Farm Bureaus: "One of the best-documented facts about the Farm Bureau Movement is that it was a direct outgrowth of the County Agent System. All the instructions to agents from their department were based upon the assumption that the first part of the agent's work would consist in organizing Farm Bureaus" (McConnell, 1953:48). In fact, in California, the university actively discouraged the formation of other agricultural organizations that might compete with the Farm Bureau (Fiske, 1979). This appeared reasonable to college officials convinced of the "unorganized" character of rural life. Indeed,

> since contributions from private sources were specifically recognized in the Smith-Lever Act of 1914 as legitimate sources of state matching funds, in some areas county appropriations and farm bureau membership dues came to make up a substantial portion of the funds used for the county agent's salary and his work. Thus the informal impulse for the promotion of farm bureaus by the Extension Service was reinforced in some cases by legal ties (Campbell, 1962:6).

By 1920, the county Farm Bureaus had established a national federation, and the Association of Agricultural Colleges was concerned about its close alignment with the Farm Bureaus, which some college officials saw as political and not educational organizations. Congress launched an investigation and concluded that the American Farm Bureau Federation could not be both a lobby and a quasi-governmental organization. In 1921, A. C. True reached an agreement clearly distinguishing the county agents from the Farm Bureaus. "It was definitely recognized that the extension forces should confine their activities to educational work and should not become organizational agencies for any association" (Clinton et al., 1921:223). Nevertheless, an extremely powerful lobby that shared with scientists the ideals of organization and efficiency (Danbom, 1979) had been established at public expense.

Once established, the local Farm Bureaus appeared to provide an effective way of utilizing limited extension staff. E. D. Merrill, dean of the University of California, expressed this efficiency as follows: "Working with individuals is, however, time consuming. Working with or through organizations may mean multiplying manyfold the results of effort put forth by experiment station or extension men. For this reason there has been a particular interest . . . in affiliation with organized farmers" (1928:101). At the same time, as Merrill recognized, such cooperation with actively political organizations raised a number of problems for the land-grant schools. During the mid-1930s, American Farm Bureau Federation (AFBF) membership increased dramatically, and the AFBF developed a cordial relationship with New Deal officials. Despite official assurances of separation, Edward A. O'Neal, president of the AFBF, could boast in 1936 that "in Illinois there are many counties where the Farm Bureau in its own building, houses the activities of practically every agricultural agency in the county" (1936:49). Increasingly, however, the AFBF saw USDA as "created for the purpose of serving the farmers alone" (Campbell, 1962:160). While economists wished to reform the agricultural sector, the Farm Bureau was concerned almost exclusively with price policy. The economists' action programs, supported by AFBF's much smaller competitor, the Farmers Union, were seen not merely as help for others but as a direct threat to the AFBF. As a result, the AFBF sought and successfully accomplished the dismissal of reform-minded social scientists and the dismantling of the Bureau of Agricultural Economics (Kirkendall, 1966). Only in recent years has the successor to the BAE, the Economic Research Service (ERS), begun to address broad agricultural issues.

Commodity Groups

Over the years, it became apparent that there was no *one* agricultural science. Gradually, the disciplines developed, and research became increasingly commodity oriented. This commodity focus was due in part to the scientists' training, which stressed a reductionist perspective, as well as to the orientation of the prosperous farmers who supported their research. Unlike their poorer contemporaries, wealthy farmers often specialized in the production of one or two commodities and organized themselves into national commodity associations in order to increase their influence on the legislative process. This practice was part of a general shift from local to national interest groups occasioned by the introduction of science into politics (Hays, 1959). These associations have lobbied for the continued development of commodity-specific research. For example, Moore informs us that "An organization of cotton growers and processors has been one of the most consistent supporters of ARS [Agricultural Research Service] research and its most effective lobbyist for increased funds. Other groups lobby for other commodities. A group representing flour millers, transportation, and the grain trade has been very effective in getting congressional support for research on grains . . ." (1967:146). Similarly, at the state level, Fujimoto and Kopper (1975) have found California marketing orders (i.e., organized commodity associations) to be particularly influential in lobbying for commodity-specific research in that state.

This commodity focus extends to the various USDA and SAES advisory committees. As Mainzer noted some years ago, when these committees were established, "the 'little man' is seldom chosen, for the organization leaders and those that are best known within their industry and within the department are large operators" (1958:315). Recently, Meier estimated that the Agricultural Research Service alone was supported by 117 interest groups (1978:484).

The commodity focus has also had several unintended consequences. Most obviously, it has underrepresented the smaller, mixed farmer who, by virtue of the nature of his operation, has neither time, money, nor inclination to become involved in a variety of commodity associations. Second, it has resulted in the underfunding of non-commodity disciplines and research topics (e.g., nutrition and soil conservation). Third, it has made it particularly difficult to develop an adequate overview of the whole (i.e., to develop improved farming systems). Fourth, on occasion it has brought about direct interference with research when a particular interest group felt its needs were not being served. For example, Fiske (1978) has described how sociological investigations in the 1940s ceased in California as a result of commodity group pressures. In a later work, he notes that a wildlife biologist was dismissed from the California experiment station in 1964 as a result of his statements about pesticide use (Fiske, 1979). As Hardin put it several decades ago, "The writer knows of no college of agriculture in which some professor has not been subjected to pressure—attempts to get him fired, to silence him on an issue, to force retraction of a publication, to require that a controversial manuscript be reviewed by representatives of an affected interest, or simply protest enough so that he will think twice before he repeats the 'offense' " (1955:86).

And, alas, nearly twenty years later Breimyer reported that such "instances of gross misconduct" had not ceased (1973:995).

Perhaps, most subtly, the commodity orientation has tended to direct researchers and administrators away from controversial areas (Hardin, 1955) and toward those that would not disturb their narrowly defined and increasingly highly differentiated clientele. Over the years, this close relationship between researchers and commodity groups has come to appear so "natural" that scientists rarely question it (McCalla, 1977). Even those who raise the relationship as an issue are often viewed with hostility (Hadwiger, 1975).

More recently, the commodity groups have begun to overshadow more broadly based farm organizations. As James H. Ebbinghaus of the Farm and Industrial Equipment Institute has noted: "The individual farmer's concern has been narrowing from broad farm policy to specific policies that affect his particular product. . . . I submit that special interest groups in agriculture is where the 'action is!' " (1978:109–110; see also Bonnen, 1967).

This shift has helped to alter radically the nature of national farm politics. The days when agricultural bills would sail through Congress have ended. "In the late 1960s, the commodity interests fragmented making it more difficult to reach agreement among the farm block [*sic*]. Labor and consumer groups began asserting a crucial role in determining farm policy; yet even as late as 1973, farm bills sped through the House in a day. Now, if they pass at all, a week is the minimum time for a bill to clear" (*Agricultural Engineering*, 1979:15). During the fall of 1981, many of the commodity bills were passed with great difficulty. Continued fragmentation is likely to make passage of farm bills more difficult in the future.

Agribusiness

Another group of clients, agribusiness, consists of the complex of corporations supplying agricultural inputs, processing outputs, and, more recently, vertically integrated into the respective commodity production systems developed somewhat later. Unlike even the wealthiest farmer, these agribusiness corporations have been able to provide funds to experiment station researchers for specific projects. An early survey commented: "Of 41 institutions reporting on the question, 15 had no agricultural research in cooperation with commercial agencies in 1928. Twenty-six institutions, however, list for that year a total of 120 cooperative research projects, involving a total of 113 cooperating commercial agencies" (Klein, 1930: vol. 2, 602). This presented serious administrative and political problems for station directors. On the one hand, they were quite short of funds for research. On the other hand, they were well aware of the potential dangers involved in accepting commercial support. The Association meetings of 1928, 1929, and 1931 focused, in part, on this issue. As one participant summarized it: "In accepting monies from industry, that station director must always keep in mind that such monies should have as their first purpose the upbuilding of agriculture in its broadest sense. We should not use our station facilities for the promotion of merely private gain" (Russell, 1931:226). But there was a catch: "The field of the physical sciences is not likely to result in trouble because experimentation and observation are factual in character; and conclusions generally have to follow where the facts lead. In the realm of economics, the likelihood of trouble ensuing is much greater; perhaps this is the case in the social field" (Russell, 1931:226). Today, many agricultural scientists would still accept this argument. Yet, when examined more closely, this distinction between natural and social sciences proves to be fallacious. One need only look at the response of scientists to rising energy costs to see that natural science research is indeed influenced by changing economic and social conditions. Looking further, one sees that when the decision to pursue a certain line of research is made, implicit *political* decisions are made about the beneficiaries. Of course, this is not to suggest that individual scientists deliberately set out to ignore certain client groups. However, scientists *are* encouraged by both administrators and disciplinary organizations to work on "important" problems; and what is "important" is defined, in large part, by the political-economic-social system in which we live.

These examples reveal that the current structure of the agricultural sciences is built upon negotiations within those sciences and with various other individuals and interest groups. Many of these negotiations have become fully integrated into structure (e.g., the definition of a valid experiment). Others, however, reappear every so often in somewhat different form (e.g., the basic/applied debate, the relations between scientists and client groups), despite vigorous efforts to resolve them.

Variations on a Theme—Disciplinary Differences

Up to this point, we have treated the agricultural sciences as if they were homogeneous. Indeed, they are in that (1) they were all established by governmental order; (2) they all operate, in part, within the administrative structure of agricultural colleges and USDA research services; (3) they emphasize a mission orientation, a concern with serving a clientele; and (4) they have enjoyed continuous formula funding for nearly a century. On the other hand, they are sharply differentiated from each other by their different historical foundations, subject matters, and levels of funding. Let us examine each of these differences separately.

Historical Origins

In general, the agricultural sciences originated not in response to scientific requests but in response to journalists, farmers, and certain chemists impressed by the work of Liebig (Dupree, 1957). For example, "economic" (applied) entomology emerged to a large extent as a peculiar product of USDA. "Unlike such fields as soil science, which flourished in Russia, economic entomology had no equivalent abroad" (Rossiter, 1979:220). However, a link with traditional academic science was soon forged. This link was probably one of administrative convenience, for, until the middle of this century, basic entomologists and "economic" entomologists maintained two separate professional societies. In fact, economic entomologists tended to view insects as pests, but their colleagues saw insects as beneficial (Rossiter, 1979). In other disciplines, those with applied interests tended to be segregated from those with "basic" interests and maintained their own societies. This separation may have insulated the applied scientists from certain criticisms and increased their identification with client groups. In entomology today, however, there is a single entomological society which is dominated, by virtue of both numbers and dollars, by the economic entomologists. As a result, within the society serious debate continues over appropriate relationships with the pesticide manufacturers. As the president of the Entomological Society of America (ESA) recently put it, "There is a very important, if not large, number of members and potential members who feel that ESA is involved in matters not the proper concern of a scientific society" (Sailer, 1977:1).

Forestry too has its tensions between applied and basic research. Forestry has traditionally served the sometimes contradictory goals of conservation and productivity. Unlike other agricultural disciplines, forestry was governmentally funded, in part, because of public concern over conservation. Until the 1870s, Americans had seen forests as inexhaustible sources of timber. The 1871 fire in Peshtigo, Wisconsin, that killed 1,500 persons and burned over 1.3 million acres first raised the question of wilderness conservation to the status of a major issue.

Gifford Pinchot, a Yale graduate and friend of Theodore Roosevelt, became the first chief of the Division of Forestry in 1898. Although he was dismissed upon Taft's election, Pinchot left his mark on what was to become the Forest Service. He established the first forest experiment station, founded the American Society of Foresters, helped establish a graduate school of forestry at Yale University, encouraged the development of forestry at SAES, and created a certain esprit de corps within the Service. Pinchot, who saw the unmanaged use of resources as a threat to the nation (Frome, 1971:14), "brought the word 'conservation' into popular usage in application to natural resources" (Frome, 1971:12). At first, there was little opposition to the conservation theme. However, when opposition did develop, Pinchot wasted little time in bringing the issue before the public. This had the curious effect of attracting to the movement those who wished to *preserve* as well as those who wished to *conserve*. Pinchot was disturbed at this turn of events, for "his basic view [was] that the reserves should be developed for commercial use rather than preserved from it" (Hays, 1959:51). This conflict over the interpretation of "conservation" became a part of the ongoing public debates concerning forestry and a feature of forestry as a discipline.

Eventually, Pinchot's strong conservationist stance came to be challenged. His proposal for government control of cutting on private lands was not supported by his colleagues. While the western forest experiment stations had been established to do research relating to the immense public holdings, those established later, in the East, were created through the pressure brought by timber owners (Dana, 1950:320). As a result, the eastern stations were more concerned with increasing yields than those in the West.

Over the years, this tension between production and conservation has prevailed both within the Service and outside it. Internally, scientific results have occasionally been subordinated to Service policy (Schiff, 1962). Externally, "its legislative programs are followed by a wide spectrum of interests, ranging from citizen conservationists . . . to intensely commercial industries dependent upon raw materials from public lands for continuing profits" (Frome, 1971:162–163).

In recent years, forestry has been associated with social science research on leisure and recreation. Unlike almost any other journal published by a disciplinary society in agriculture, the *Journal of Forestry* publishes information on new legislation, lively critical letters to the editor, and critical state-of-the-discipline articles (Lacy and Busch, 1982b; Minckler, 1976). Recently, it has even acknowledged a problem of sexism within the discipline (Whisnant, 1976).

Agricultural engineering provides still another variant. Arising out of the mechanical work related to the construction and maintenance of farm equipment and including civil, electrical, and mechanical engineering, agricultural engineering had to struggle to obtain recognition as a distinct discipline. The first ten volumes of its major journal, *Transactions*, contain little or no research (Stewart, 1979).

The practice of refereeing journal articles began only in the early 1950s.

Ties between agricultural engineering and industry were forged early in the century. By the late 1920s, advertising by commercial firms was common in engineering journals. A very substantial proportion of ASAE members were and remain industry engineers. Agricultural engineers have tended to take an approach quite different from that of many of their colleagues. Engineer Robert E. Stewart explains: "Their basic approach was to increase per capita production through mechanization of large, efficient farms thus reducing cost of production per unit produced, just as in any American industry. This philosophy committed them to the 'mass production' farm, even though American society thought such farms were socially detrimental" (1979:93).

Agricultural engineers have apparently been more resistant than their colleagues in other disciplines to external and internal criticism. An attempt by one ASAE member in 1961 to produce a "reappraisal" of the role of agricultural engineering was apparently turned into a public relations campaign (Stewart, 1979:269). There have been some recent attempts at a serious appraisal of the discipline, but shrill voices still occasionally insist that some questions should not even be asked.

Subject Matter

The subject matter of the agricultural sciences has a substantial and differential impact on them. For example, crop agronomists and most horticulturists are tied to an annual cycle in their research. A typical agronomic experiment will take at least three years to complete. Foresters and horticulturists working with tree crops must have substantially longer temporal perspectives, as is also the case for animal scientists. On the other hand, soil agronomists, many entomologists, and agricultural engineers can often complete experiments in under three months. Such differences, of course, affect the *potential* responsiveness of the respective disciplines to the problems of client groups.

Client Groups

The agricultural sciences also serve different, though overlapping, groups of clients. Most agronomists study problems of interest to large, though generally family owned and operated, grain farms. An exception is most tobacco research, which is oriented, in general, toward small to medium-sized farms. Horticulturists, with the exception of those whose research is useful to giant California-based vegetable operations, are generally oriented toward medium to small farmers. However, they have not as yet taken seriously the impact of the recently revived backyard gardening movement (United States Congress, 1976:9). Horticulturists probably have the largest—if not the most powerful—organized constituent group of any agricultural science in the form of the various state horticultural societies. These societies, whose members are home gardeners, often predate organized horticultural science by a century. Until recently, local societies faced declining memberships and the indifference of horticultural scientists. Of late, however, their membership has increased dramatically and relations with horticultural science have improved.

Agricultural engineering and food science are, perhaps, the fields most closely aligned with agribusiness. Their disciplinary organizations have a strong industry focus, and industry advertising fills their journals. They also appear to share an overriding faith in the ability of science and technology to solve all problems associated with agriculture; an article in *Agricultural Engineering* asked the rhetorical question, "Is there any reason we should not expect the next 100 years to record the same progress as the last 100 years have provided?" (Lanham, 1976:19).

Animal scientists have more varied groups of clients, who range from the huge midwestern feedlot operators to the one-person feeder pig operations common in the South. Poultry scientists have discovered only recently that their successful promotion of large-scale poultry operations has sharply reduced the need for their services. Between 1960 and 1971 the number of academic poultry science departments declined from 44 to 22 (Lower and Quarles, 1975:955). Yet, the response to declining support has been generally more of the same. One observer suggested that we must "not lose sight of the fact that the only reason for us to exist as Poultry Science Departments is to support the growth and development of the poultry industry . . ." (Cook, 1977:3).

Veterinary scientists have an unusual position within the agricultural sciences in that their major clients are the graduates of the various veterinary schools, most of whom work as private practitioners. Entomologists might be expected to have extremely close ties to the pesticide companies; however, the nature, acceptability, and strength of those ties remain subject to heated debate within the discipline.

Some fields, among them rural sociology and soil science, have no clearly defined client group. As a result, they tend to be relatively underfunded and occupy a somewhat tenuous position—at Texas A&M the entire rural sociology department was recently disbanded and then reinstated—within the various public research organizations.

In short, far from being homogeneous, the agricultural sciences are quite diverse. The sources and nature of this diversity are traceable

FIGURE I.2
An Overview of the Worldwide Development of the Agricultural Sciences.

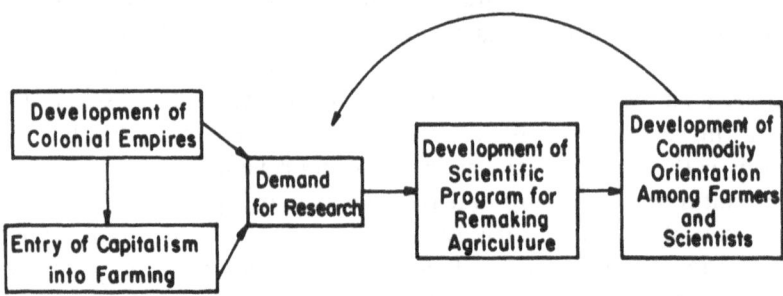

to their unique historical origins and subject matters as well as to a complex set of internal and external negotiations, which have led to the development of somewhat different structural characteristics in each discipline. These characteristics help to explain why Fujimoto and Fiske (1975:9) found wide differences across departments in the way research priorities were established at the University of California. Even this cursory description suggests that reorienting the agricultural sciences to face new problems will require: (1) programs directed to the structure and processes within agricultural research in general, and (2) strategies specifically tailored to the unique structure of each discipline.

Conclusions

Worldwide, agricultural science was developed as a part of: (1) the development of colonial empires, and (2) the shift in perspective and lifestyle from subsistence to capitalist farming (Figure 1.2). The U.S. situation differed from others in that grains rapidly became the major U.S. export crops. On the other hand, it was similar to the worldwide movement in that it depended fundamentally on the farm or plantation owners' concern for ever increasing productivity and their inability to profit individually from research that would increase productivity. Indeed, without the growers' support, no research would have occurred at all.

Like their contemporaries in other sciences and the general public, agricultural scientists in the early years of this century tended to see the products of science as undiluted good. While society might be guided by science, it seemed that science itself was autonomous—a type of knowing that, through the use of certain special methods, ensured the emergence of truth and social progress (Jordan, 1907). To this end, the leaders of the agricultural colleges felt compelled to embark upon a program for remaking rural society through the application of scientific knowledge. The increasingly specialized, commodity-specific nature of scientific inquiry encouraged the development of a commodity orientation among clients (Rossiter, 1979; Boss, 1927). Among the unanticipated consequences of this development were the underfunding and abandonment of problems not directly related to commodities. Gradually, over the course of a century, this relationship began to appear to many as "natural" (McCalla, 1977).

What scientists failed to comprehend was that science itself is a social process (Latour and Woolgar, 1979) that is guided by the shared aims of scientists, aims that are formulated in the political sphere and that are often strongly biased in favor of powerful interests (Feyerabend, 1978). The very claim to scientific autonomy had the effect of making science the pawn of vested interests because it served to divert scientific attention away from questions of ends—What is the good society?—to questions of means—How can productivity be increased? Thus, even when the products of agricultural science seemed to have deleterious consequences for the larger society, they could be dismissed, recast as adjustment problems, or simply tolerated by scientists and farmers alike as the price to be paid for inevitable scientific advance. Both the belief in the autonomy of science and the organization of influential clients along commodity lines made it possible for productivity and efficiency to be treated as ends and for scientific work to be gauged more and more by the degree to which it contributed to these "ends."

From the mid-1930s through the 1960s, rare was the voice that questioned the primacy of the quest for increased productivity. Although it was occasionally suggested that the products of science were encouraging migration from farms and that too great a decline in farm population might not be desirable, others saw this as the inevitable course of scientific advance (Stewart, 1979). In the last decade, however, new questions have been raised, among them: Is current U.S. agricultural productivity sustainable? Is extreme regional specialization in agricultural production desirable? Can we affort such an energy-intensive agriculture? Can we continue to apply large quantities of chemicals without causing various kinds of environmental problems? What are the consequences of food and agricultural policies and practices for human nutrition? for quality of life? These and other issues will not be resolved easily, but they have the combined effect of focusing attention on ends rather than on means and of suggesting the need for an examination of the research process itself. In the research reported here, we begin a careful study of the U.S. agricultural sciences in the hope of contributing to the discussions among both scientists (e.g., Cheney, 1974) and the public at large about the role and direction of the agricultural sciences.

2
Organizational Structure and Problem Choice in Agricultural Research

Organizational Structure

The structure of research typically found within publicly funded U.S. agricultural research systems today is highly complex (Figure 2.1). At the center of this system are the agricultural scientists themselves, including both the natural and social scientists (2 and 3, Figure 2.1). Each of the natural and social sciences can be broken down into various research groups working on particular research problems. Each research group consists of one or more research scientists, technicians, and assistants being trained in that particular science. These groups draw upon knowledge previously constructed by their colleagues and reported by other scientists. Thus, they overlap to some degree with one or more academic client groups.

Closely linked to the scientists are various clients, users and other system participants (1, Figure 2.1). These include operatives or practitioners (farmers and others who directly employ the research products). They also include intellectuals who integrate various research findings into broader theoretical frameworks than those found within individual sciences (e.g. Commoner, 1972; Land, 1973). A third group are private funding sources (e.g., the manufacturers of farm inputs). A fourth group are "publics" (e.g., agribusiness firms, multinational corporations, foundations, government agencies, and homemakers), who employ and/or request the products of certain types of natural and applied social science research. A fifth group of system participants are administrators within the college or research service who make decisions about the use of public funds. Sixth are "other scientists." They are distinguishable from colleagues in that they are not engaged in the same sub-specialty, although they employ the results of research

38

FIGURE 2.1
Schematic Representation of the Agricultural Science Enterprise[a]

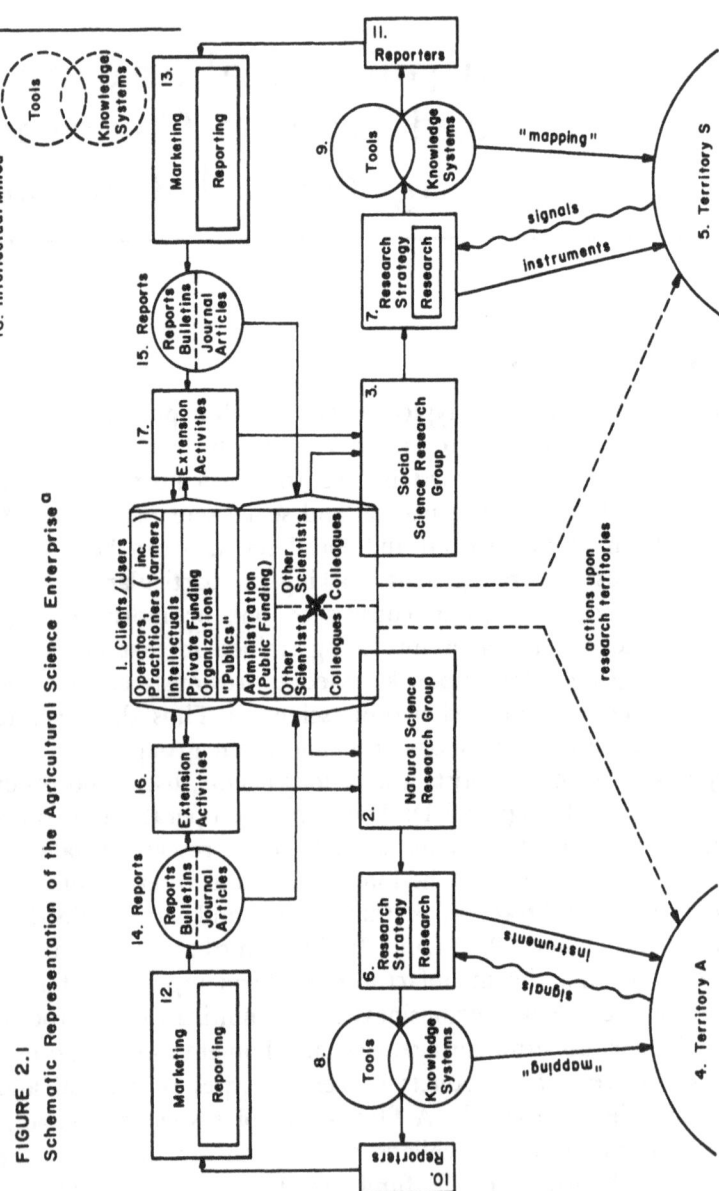

[a]Adopted from Radnitzky (1973:1)

Used by Permission of Humanities Press Inc., Atlantic Highlands, New Jersey 07716.

in their respective fields. The double arrows between "other scientists" and "colleagues" are meant to indicate that "colleagues" within the social sciences are "other scientists" to those within a natural science and vice versa. These groups often overlap.

Each group of system participants presses its demand for a particular type of research upon the various research groups (Fujimoto and Fiske, 1975; Fujimoto and Kopper, 1975). It may do this directly through interaction with researchers, by providing funds through legislative pressures, and/or through extension staff (16, 17, Figure 2.1). For example, cotton growers have successfully lobbied for public funds earmarked for cotton research; in contrast, many pesticide manufacturers provide small private grants to university researchers to test new chemicals. Each research problem that is acted upon reflects, in part, the varying influence of each group within the system and is the result of a rather complex and continuing process of negotiation (Strauss, 1978). The different groups act upon the research territories (4, 5, Figure 2.1) of both the natural and social sciences. Research territories are the "objects" with which a research group is concerned. They may be geographical, but more often they are *abstracted* aspects of certain phenomena (a corn virus, youth, pork belly futures). System participant actions may be primarily technical— for example, the widespread application of a particular pesticide. Alternatively, they may be primarily social, as in the reorganization of an agricultural marketing system. In addition, under certain circumstances, client groups may serve to define new territories of study or to redefine old ones. For example, Perkins (1977) showed how several social crises forced entomologists to redefine pest control strategies.

Under conditions of "normal" science (Kuhn, 1970), research groups in both the natural and the social sciences select particular research strategies (6, 7, Figure 2.1), such as field trials or survey questionnaires, that are scientific traditions for dealing with particular types of research problems. These strategies typically suggest the employment of certain kinds of instruments for dealing with the respective research territories (4, 5, Figure 2.1). What categorizes all instruments is that they are ways of very selectively acting upon a research territory so as to receive certain expected signals. For example, a laboratory scientist might employ certain types of laboratory equipment so as to obtain desired meter readings; a social scientist might employ a survey questionnaire so as to receive checked boxes on a survey; and an agronomist might apply fertilizers in varying dosages on given fields so as to obtain data on crop yields. These instruments, though themselves technical accomplishments of substantial import, are of no more than historical interest to scientists, much as the *development* of research products is rarely of interest to clients and users. Particular instruments are expected to yield particular signals when employed in their appropriate research territories. They *do not* provide "copies" or "pictures" of what is studied, but instead report upon certain carefully selected features of the research territory. Unfortunately, despite its abandonment elsewhere, many agricultural scientists still cling to the "copy theory of knowledge" (Lanham, 1976).

The signals received are then interpreted by the researchers in light of their expectations (i.e., hypotheses are tested). As a result, new tools and knowledge systems are developed (8, 9, Figure 2.1). The tools are new ways of dealing with certain aspects of the world (e.g., a new fertilizer or social policy) while the knowledge systems may be regarded as "maps" of aspects of particular research territories. Knowledge systems are typically explicated in textbooks, and as they are modified through research, the textbooks are rewritten (Kuhn, 1970). Tools and knowledge systems overlap in the sense that certain tools are meant for use with particular knowledge systems. For example, a new social policy is meant for use within the sociological knowledge system.

Researchers, then, act as reporters (10, 11, Figure 2.1) and market their research (12, 13, Figure 2.1) in reports (14, 15, Figure 2.1). These reports typically take the form of papers, journal articles, or books, which are then made available to client groups. In addition, the results of public research are frequently published in an array of in-house "bulletins." These publications generally report location-specific research results. In this sense, the agricultural sciences differ substantially from their "basic" science counterparts. Indeed, the location-specific character of agricultural research has led some critics to believe erroneously that it is unnecessarily repetitive. It is ironic that this ecological feature—the need to test a new product in a variety of niches—is only now, after nearly a century of research, beginning to enter into the sciences themselves in the form of intercropping and farming systems research.

Additional marketing and reporting are done by the Cooperative Extension Service. The (federal and state) Extension Service (16, 17, Figure 2.1) translates the highly technical reports of scientists into languages more readily understood by technicians and various "publics." In some cases, this translation involves two different persons; in others, it requires the researcher to "change hats" and act in an extension capacity. States vary widely in the degree of distinction made between research and extension. The Extension Service, and scientists in their extension roles, also serve as a source of research problems by transferring and translating lay research requests into the technical language of a particular research group. For example, "Why has my corn died?" must be translated into, "What new type of leaf rust has appeared and what chemical compound or genetic change is necessary to control or eliminate it?" The translation process is of particular importance in that the formulation of a research problem limits the range of acceptable solutions. Indeed, in the case of entomology, the emphasis on chemical control in the 1930s and 40s may have delayed work on alternative strategies.

Finally, the entire process of scientific research takes place within an intellectual and political milieu (18, Figure 2.1) in which certain tools and knowledge systems are taken for granted. For example, a sociologist who employs mail questionnaires assumes that the post office will deliver the questionnaires and that the potential respondents will be competent to read them. Similarly, the entomologist who wishes to develop a new pesticide assumes that the basic chemicals and laboratory apparatus will be produced and distributed in such a way as to make them available. The knowledge system also includes an assumed understanding of the goals of research and the uses to which research will be put. This milieu is background, called into question only when something goes wrong—for example, when a needed piece of apparatus is not available, when energy costs suddenly double, or when budgets are slashed.

Criteria for Problem Choice

The broad organizational structure of the agricultural sciences provides the context in which key decisions and choices are negotiated. One of the critical decisions for these sciences, as well as for scientists' careers, is the selection of research problems. However, little systematic work has been conducted to explore this process. Recently, researchers studying science have begun to adopt the position that problem choice must be a central issue in studies of scientific development (Busch and Lacy, 1981; Edge and Mulkay, 1976; Gieryn, 1978; Weinstein, 1976; Zuckerman, 1978). The current investigations of research problem choice have taken various approaches.

Some researchers have examined problem choice in the sciences by exploring how problems become defined as interesting or even as basic. Zuckerman, in summarizing some of these studies, concluded "that scientists define some problems as pertinent and others as uninteresting or even illegitimate, primarily on the basis of theoretical commitments and other assumption structures" (1978:74). Theory and its associate concepts can preempt research attention by defining certain observations as irrelevant, specifying certain investigations as unfeasible, defining certain areas as not problematic, and directing attention from certain issues. As Kenneth Burke observed, "a way of seeing is also a way of not seeing—a focus upon object A involves a neglect of object B" (quoted in Zuckerman, 1978:81).

Another approach has been to focus primarily on the emergence of science specialties and to treat problem choice as a collateral issue. The background, social characteristics, and research experiences of scientists are examined to discover patterns among scientists entering certain fields or specialties (Edge and Mulkay, 1976). Related work focuses on identifying sequences of change and continuity in the problem choices of scientists, and recognizing the social and cognitive conditions that contribute to continuity for certain research problems and those conditions that lead to problem change (Gieryn, 1978).

A third equally important research perspective focuses on how scientists choose from the range of identified problems. This orientation has explored the determinants or criteria for problem selection (Busch, Lacy, and Sachs, forthcoming; Edge and Mulkay, 1976; Lacy, Busch, and Sachs, 1980; Zuckerman, 1978). Several criteria have been suggested as the major determinants for problem choice. Zuckerman (1978) concluded that two main criteria were: (1) the assessed scientific importance of a problem, and (2) the feasibility of arriving at solutions. The importance of avoiding error-prone fields and focusing on soluble research problems was stressed by Medawar:

> No scientist is admired for failing in the attempt to solve problems that lie beyond his competence. The most he can hope for is the kindly contempt earned by the Utopian politician. If politics is the art of the possible, research is surely the art of the soluble. . . . *Good scientists study the most important problems they think they can solve.* It is, after all, their professional business to solve problems, not merely to grapple with them. The spectacle of a scientist locked in combat with the forces of ignorance is not an inspiring one if, in the outcome, the scientist is routed. *That is why some of the most important biological problems have not yet appeared on the agenda of practical research* (1967:7; emphasis added).

The interplay of scientific or paradigmatic criteria and social factors in problem choice is complicated, as judgments about appropriate research are also influenced by social processes internal to science. Merton noted that research problem choice may be influenced by "reactions to the inferred critical attitudes or actual criticism of other scientists and by an adjustment of behavior in accordance with these attitudes" ([1938] 1970:219). In addition, the social system of science provides institutionalized motivation and reward not only for solving major scientific problems, but for solving them first. This reward system motivates large numbers of scientists to migrate to interesting and substantively important emerging areas, often referred to as "hot topics." Sullivan, White, and Barboni (1977) noted that physicists try to maximize the chances of both achieving priority and solving significant problems. This emphasis on priority may increase the motivation of scientists to choose research that has a high probability of publication in professional journals.

Other researchers have raised the issue of important extrascientific influences on problem choice. Merton ([1938]1970), in his analysis of science in seventeenth-century England, concluded that research problem choice followed directly from intrinsic scientific and technical developments and indirectly from scientific concern with extrinsic military, economic, and technological problems. Weinstein (1976) argued more forcefully that the intrinsic scientific or paradigm-related criteria constitute only one of four important groups of criteria for problem choice. The other three determinants are administrative directive, political commitment, and personal avocation. Administrative directive refers to standards external to the scientific discipline or institution, but relevant to some other public or private institution. Some argue that the key criteria for problem choice may simply be the research agendas established by those institutions providing the funding. For example, C. Wright Mills stated that since many studies in the social sciences are "quite expensive, they have had to be shaped by some concern for the problems of the interests that have paid for them" (1959:64). Similarly, Fujimoto and Kopper (1975) and Evenson and Kislev (1975) proposed that the most important external influences upon agricultural scientists' research choices are various commodity groups that financially and politically support selected aspects of agricultural research.

Frequently, the alternatives suggested by those who accuse scientists of prostituting themselves for grants and other benefits are equally nonscientific criteria that involve political commitment. Weinstein (1976) noted that even the natural sciences may use criteria for problem selection that involve some political values. For example, a biologist may concentrate in the sub-field of ecology out of a value commitment to improving the quality of the environment. Similarly, a geneticist may refrain from certain types of research because the results could be used for immoral purposes by certain powerful groups. Finally, Weinstein notes that a scientist may choose a research problem for personal reasons such as indulging psychological quirks or turning a hobby into a professional service. We noted in an earlier paper

(Busch and Lacy, 1981) that a scientist's nonscientific interests and life-style may influence research problem choices. During an interview, one agricultural scientist who engaged in a great deal of field research indicated that his choices were greatly influenced by his desire to work outside. A plant pathologist reported that his research interests developed from his hobby of ornamental horticulture.

Among researchers studying science, a major debate has emerged regarding which of these criteria operate in the sciences. This brief summary suggests that the scientists' choices of research are influenced in a wide variety of ways by a diverse array of factors. Some of these criteria are clearly disciplinary in character, while others are external to science. It is often difficult to make a clear distinction between internal and external influences. Furthermore, two or more criteria are frequently operating in the choice of research problems, and it is difficult to judge which is the most important, or even which ones are present. Decisions made by scientists regarding problem choice emerge from a complex process of negotiation within themselves and with other scientists, administrators, and clients (Busch, 1980).

To examine the criteria used by agricultural scientists in this complex process, we began by identifying the criteria suggested in the literature. We then conducted a series of semistructured interviews with key informants in the fields of agronomy, agricultural economics, agricultural engineering, animal science, biochemistry, entomology, food science, forestry, genetics, horticulture, nutrition, phytopathology, and rural sociology (Busch and Lacy, 1981). From this series of in-depth interviews, we developed a relatively comprehensive list of 21 criteria for research problem choice. Finally, a random sample of scientists listed in the Current Research Information System (N=1,431) were surveyed by mail about the importance of these 21 criteria for their choice of research problems (Table 2.1). The sampling procedure is described in the Methodological Appendix.

"Enjoyed doing this kind of research" emerges as the single most important criterion for research problem choice. The second most important criterion is "importance to society." However, personal interviews and correlation analysis suggest that this criterion glosses over a rather heterogeneous group of meanings. Specifically, "importance to society" is most highly correlated with the "likelihood of clear empirical results" and the "marketability of the final product." On the other hand, although scientists rate "importance to society" highly as a criterion for problem choice, they do not see "community improvement" or "improved level of living" as important research goals (Chapter 9). The third item in Table 2.1 underscores the importance of available facilities to the conduct of the research process.

TABLE 2.1
Criteria for Research Problem Choice

Rank	Criterion	Mean Score [a]
1	Enjoy doing this kind of research	5.86 [n]
2	Importance to society	5.59 [o]
3	Availability of research facilities	5.24 [p]
4	Scientific curiosity	5.13 [pq]
5	Potential creation of new methods, useful materials, and devices	5.03 [qr]
6	Publication probability in professional journals .	4.89 [rs]
7	Client needs as assessed by you	4.87 [rs]
8	Likelihood of clear empirical results	4.76 [s]
9	Funding	4.75 [s]
10	Evaluation of research by scientists in your field	4.34 [t]
11	Priorities of the research organization	4.29 [t]
12	Potential contribution to scientific theory	4.26 [tu]
13	Demands raised by clientele	4.10 [uv]
14	Credibility of other investigators doing similar research	3.95 [vw]
15	Currently a "hot" topic	3.84 [wx]
16	Length of time required to complete the research	3.79 [x]
17	Potential marketability of the final product . . .	3.74 [x]
18	Colleagues' approval	3.74 [x]
19	Publication probability in experiment station or research service bulletins and reports	3.64 [x]
20	Feedback from extension personnel	3.49 [y]
21	Publication probability in farm and/or industry journals	2.98 [z]

[a] Mean score based on seven point scale (1 = not important, 7 = very important).

[n-z] Criteria with different letters are significantly different, $p < .01$. (See Methodological Appendix for details.)

Clearly, no research can be undertaken without space, equipment, libraries, and other paraphernalia of research.

Scientific curiosity ranks fourth among all scientists in our list of criteria. As one respondent rhetorically asked, "The scientists' dream is science for the sake of science. Almost everything and everyone a scientist has to deal with is the antithesis of this. Who is wrong?" The potential creation of new methods, useful materials, and devices is ranked fifth. Clearly, many agricultural scientists see the development of concrete products as a central feature of their research.

Not surprisingly, publication probability in professional journals is rated high on the list of criteria. In contrast, publication probability in bulletins and in farm and/or industry journals rank 19 and 21. One scientist pointed out that the emphasis on journal publication contributed to increasing specialization in the agricultural sciences:

> I am all too familiar with theoretical researchers who have played the "publish or perish" syndrome to the hilt, but have not contributed one bushel increase to crop yields. These people, usually department heads or location leaders, have related almost none of their papers to the consuming public. The "generalist" is almost an extinct species, and we will regret this situation in the future.

While bulletins and farm journal publications are not important criteria for research problem choice, Vacin pointed out that they are very important sources of information for farmers: "Information provided by land-grant university sources were [*sic*] read thoroughly by an average of 39.5% . . . of the readers surveyed" (1979:i; see also Brennan, 1975).

The importance of communication with clients in the choice of scientific research problem is reflected in three items ranked 7, 13, and 20. Of particular note here is that scientists' personal assessments of client needs are considered substantially more important than either demands raised by the clients themselves, or feedback from extension staff. (For an extension view see Sprott, 1978). Of perhaps even greater importance is the strong positive relationship between the percentage of one's time allotted to extension activities and the relative ranking given to the above items. Feedback from clients and extension staff appears most effective when researchers devote a portion of their time to extension activities.

These rankings suggest that formal feedback mechanisms are relatively ineffective in directing research. With somewhat less clarity, they also suggest that scientists are likely to rely on a common sense assessment of client needs rather than on a careful inquiry into the problems experienced by various client groups. Consider, for example, the words of one researcher:

> Researchers in agricultural economics (as in most disciplines, I suppose) have difficulty in determining what research would be most useful. They prefer to research those areas in which a lot of people would appreciate getting the results. The public and their other clients, however, do not communicate their needs well so the researcher has to decide on his own what is important. As a result there is often little interest in what he does. Competitive grants should help overcome this problem in part. Selling research to private clients at least assures that someone is interested in what is being done.

An alternative explanation would be that scientists tend to take a broad, long-run view of research needs. However, many scientists complain of the short-term character of most current research:

> Short term technological solutions only put off (delay) the immutable natural consequences. In general, the research in my field is short sighted in its attempt to increase productivity and the farmers' (and others') profit. As in most ecological systems an equilibrium solution is the only one which persists.

Importantly, virtually none of our respondents are concerned that long-term research is overemphasized.

Funding ranks ninth among the 21 criteria. Several respondents feel that funding and, particularly, the search for extramural grants are greatly overemphasized:

> I strongly feel that so much attention has been paid to funding, to grants, and contracts, that the fundamental purpose of scientific research—to increase organized knowledge and how to use it—has suffered greatly. I do not hear colleagues talk of break-throughs or strong inferences anymore, only about "this wonderful new program" or the "number of Ph.D.'s we have turned out" or "how we plan to expand into this or that." The specific objective of the research is seldom mentioned, and when it is, it sounds like more of the same old language: "Improve our understanding of . . . ," "study the relationship between . . . ," and "build a model to account for . . . . " All vague, all open-ended—not a specific goal in sight. *Program funding* has become the objective of scientific research! (emphasis in original)

Priorities of the research organization rank eleventh among the 21 criteria. As one respondent put it, "most research in agricultural experiment stations is done by individuals with almost *no* directions from directors of stations. There is no priority 'plan' for [name of state]. The department head is influential, however" (emphasis in original).

Most scientists put the fact that subjects are "hot topics" relatively low on their list of criteria for problem choice. Although a small group sees this as an important criterion for problem choice, the majority of scientists appear to take a position similar to this:

> I am interested in exploring the principles of important reactions and processes that are associated with soils. My aim is to test hypotheses that I have formulated, regardless of how unpopular they may be. I believe that, in so doing, the knowledge I accumulate will ultimately benefit my discipline and mankind. Therefore, I am not interested in pursuing topics that are currently hot or that will receive generous financial support unless they coincide with my interests and goals.

Finally, as we suggested earlier, these criteria emerge from a complex process of negotiation and often operate together to shape research problem choice. As one scientist observed,

> My choice of research areas is based on a mix of my perception of important limiting factors in yield and quality of my crops, my abilities and interests, availability of facilities and type of facilities, and availability of extramural funding that is consistent with the above.

These criteria for research decisions have important implications for the processes and products of agricultural science. The next five chapters begin a closer examination of the agricultural research system: its human resources, its communication network, its academic disciplinary organization, and its organizational and extraorganizational environments. Each of these important dimensions of the research system influences the criteria for problem choice and plays an integral role in the complex way in which these multiple factors converge to construct the social product we call science.

3
Agricultural Scientists:
The Human Resource

It has been posited that "human resources are the basic determinants of the rate of development of science, technology and social institutions" (Harmon, 1965:iii). These human resources, in the form of scientists, technicians, and research associates, provide the expertise and knowledge for scientific development. They bring a variety of personal qualities, background characteristics, and social histories to the research setting. A growing body of literature and empirical studies has demonstrated the remarkable influence of these person-bound factors throughout the production, interpretation, and evaluation of scientific results (Crane, 1969; Feyerabend, 1975; Latour and Woolgar, 1979; Mitroff, 1974; Rosenthal, 1966).

However, comparatively little is known about the background, education, and career patterns of these persons, particularly in the agricultural sciences. An understanding of the research enterprise in the agricultural sciences and its ability to respond effectively to new challenges requires an examination of the scientists themselves. This chapter presents a demographic overview of the contemporary population of agricultural scientists—their family background, their education, and the general nature of their current work.

The agricultural scientific enterprise has grown dramatically during the twentieth century. This rapid growth has paralleled the great increase in scientists around the world. Indeed, it has been estimated that 90% of all scientists who ever lived are alive today. Although an ambiguity of definition produces variable estimates for the size of the world scientific community, the 1981 Statistical Yearbook edited by UNESCO (1981) reported about 3.7 million scientists in 85 countries in research and experimental development. If consideration is limited to scientists and technologists who make substantial contributions to knowledge, the number is roughly a quarter of a million (Price and Gursey, 1975).

The United States has a greater number of research and development scientists and engineers and a higher proportion of these professionals in the labor force than any country except the Soviet Union. In 1980 there were an estimated 659,000 full-time-equivalent scientists and engineers active in research and development in the United States. In addition, the National Research Council (NRC) (National Research Council, Commission on Human Resources, 1976) has approximated the U.S. population of doctoral scientists and engineers in 1975 to be 279,000. That number continues to grow. Between 1920 and 1974, the total U.S. population approximately doubled, increasing from 105 million to over 210 million. During that same period, the Ph.D. scientist population increased by a factor of 50 (Harmon, 1978).

The growth in the number of U.S. agricultural scientists with doctorates has been equally dramatic. During the five-year period 1920–1924, there were 136 doctorates awarded in the agricultural sciences. By the 1950s, the number of doctorates being awarded yearly was over 300; and during the 1970s, over 1,000 doctorates in the agricultural sciences were awarded yearly (Harmon, 1978). Nearly one-third (32.4%) of all the 16,284 doctorates awarded in the agricultural sciences from 1920 to 1974 were awarded between 1970 and 1974; over half (53.2%) were awarded during the last 10 years of the period (1965–1974).

Demographic Characteristics

Sex

Given this remarkable growth in the number of scientists and engineers, it is surprising that so little attention has been directed to exploring who these scientists are and what they do. An examination of their demographic characteristics reveals that generally women remain underrepresented in these jobs despite their rapid employment growth in recent years. In 1978 women still represented only about 9% of all employed scientists and engineers, but accounted for over 40% of all employed professional and technical workers (National Science Board, 1981). This pattern is even more pronounced among agricultural scientists. In 1976, 99.6% of all agricultural scientists were male (U.S. Department of Commerce, Bureau of the Census, 1978). All the sciences were at least 72% male (e.g., biologists, 82.2%; medical scientists, 72.8%; chemists, 90.6%), but the percentage of men was highest for the agricultural sciences and engineering. In our sample of public sector agricultural scientists listed in the USDA's Current Research Information System (CRIS), a little more than 4% are females. Although this represents an apparent increase in the percentage of women in agricultural sciences, it is primarily the result of the concentration of women in nutrition (41%), social sciences (19%), and food science (10%). In agricultural economics, agronomy, animal science, agricultural engineering, forestry, and horticulture, 1% or fewer are women. The small overall percentage of women prevented any further analysis of sex differences.

Race

Employment of racial minority groups in science and engineering has been increasing faster than employment of white scientists. However, racial minorities such as blacks remain underrepresented in these fields. In 1978 only 4.5% of employed scientists and engineers were members of racial minority groups, compared to about 9.0% of all professional and technical workers (National Science Board, 1981). Although other scientific fields are dominated by whites (physical scientists, 94.3%; biologists, 94.4%; mathematicians, 93.7%), the agricultural sciences have the highest percentage of whites (98.6%) (U.S. Department of Commerce, Bureau of the Census, 1978).

Despite the federal commitment beginning in 1890 to establish additional research and educational institutions in the South to serve black farmers, the number of black agricultural scientists has remained small. Although there are 17 predominantly black land-grant institutions in the publicly funded agricultural research system, they are considerably smaller than the State Agricultural Experiment Stations, often lack a critical mass of scientists in any one area, and are generally not research oriented. Equally problematic is the tendency for these institutions to treat research funds as "soft" money. As a result, many researchers in the so-called 1890 schools are hired on short-term contracts rather than for long-term, tenure track positions. Except at Tuskegee Institute, few research funds were available at the 1890 schools until 1967. In that year, a total of $283,000 was made available to them in funds for research (Mayberry, 1977). By 1978, annual expenditures had reached $19.3 million. Moreover, unlike the "1862" institutions, which receive significant nonfederal funds, the "1890" colleges receive 99% of their funds for agricultural research from federal appropriations. Of the 11,390 scientist-years reported in the inventory of public sector agricultural research in 1978, only 143 (1.2%) were reported at these predominantly black institutions (USDA, Science and Education Administration, 1981). This figure is consistent with our finding that only 1% of our 1979 sample of scientists are at these locations.

Age

Recently, concern over the age distribution of scientists has emerged, in part from concern over research vitality (National Academy of Sciences, 1979). An examination of· data on age reveals that the scientific community is aging. In educational institutions, the proportion of doctoral scientists and engineers under 35 years of age has declined from 27% in 1973 to 19% in 1979. The median age for all persons in science and engineering in 1978 was 43.7 years (U.S. Department of Commerce, Bureau of the Census, 1978). However, age distributions were not uniform across fields of science. The groups with the lowest median age were computer specialists (36.9 years) and oceanographers (38.6), while the highest median ages were for agricultural scientists (45.3), engineers (45.8), and earth scientists (46.0). In addition, the percentage of agricultural scientists over 50 years of age (35.2%) was again slightly higher than the percentage of all U.S. scientists and engineers over 50 (31.7%).

In our sample, the mean age of U.S. public sector agricultural scientists is 48. There is little difference among the various agricultural disciplines; only the environmental scientists have a mean age under 45 (43.1 years), and only the horticulturists have a mean age over 50. Finally, the median age of agricultural scientists appears to be increasing slowly and paralleling the increasing median age in the population. This phenomenon may continue in the next few decades, particularly if, as predicted, permanent employment opportunities continue to decline in universities and state and federal experiment stations (National Science Board, 1981).

What are the potential consequences of an aging scientific community for science? Numerous scientists, historians, and philosophers have proposed that age affects the alacrity with which scientists accept new ideas and modifications of the theories and conceptual frameworks of their disciplines (Cole and Cole, 1973; Darwin, [1859] 1966; Feyerabend, 1975; Kuhn, 1970; National Academy of Sciences, 1979). This notion, often referred to as Planck's Principle, assumes that a "new scientific truth does not triumph by convincing its opponents and making them see the light, but rather because its opponents eventually die and a new generation grows up that is familiar with it" (Planck, 1950:33–34).

T. H. Huxley was so adamant about older scientists' inability to change their minds that he exclaimed that men of science should be strangled on their sixtieth birthdays "lest age should harden them against the reception of new truths, and make them into clogs upon progress, the worse, in proportion to the influence they had deservedly won" (Huxley, 1901:117).

Hagstrom has further suggested that

> young scientists may find it easier to accept new views than old scientists, who may be more strongly committed to the earlier views. . . . Firm commitments to a theory may be achieved only by those who have used it to account for things previously inexplicable, who have experienced the range of its power and the difficulties of subjecting it to test. . . . Younger scientists . . . will perceive most clearly the incompatibility of innovations and existing theory (1965:283–284).

In a test of this hypothesis, Hull, Tessner, and Diamond (1978) examined the Darwinian revolution and found that the average age of accepters before 1869 was 39.6 years and that of rejecters was 48.1 years ($p < 0.05$). They concluded that, although age explained only 6% of the variation in acceptance, it is a significant factor in acceptance of scientific theories.

In our survey of agricultural scientists, age is also related to various criteria for problem choice. Reflecting the hypothesized speedier acceptance of new ideas, younger scientists are more likely to choose a research problem if it is a "hot topic." In addition, younger scientists are more likely to base their choices on publication probability in professional journals or experiment station bulletins, indicating the perceived need to publish in order to advance one's career.

Family Origins

In addressing the family origins of the agricultural scientists and their possible consequences for the research system, we used the occupation of the scientists' fathers as an indicator of family origins. Comparative data from the research of Harmon (1965), who surveyed a sample of 10,000 doctorates graduating from U.S. universities between 1935 and 1960, revealed that all scientists were more likely than the general population to have fathers who were professionals or executives and managers. Forty-nine percent of Harmon's agricultural scientists reported that their fathers were either farmers or farm managers in contrast to only 15.6% for all fields of science and engineering. However, in both cases, scientists were far more likely to come from a farm background than would be expected by chance.

Interestingly, our survey of agricultural scientists nearly 20 years later also reflects the strong influence of a farm experience. Although the U.S. farm population is now less that 4% of the total population, 38% of all agricultural scientists come from farm backgrounds (Table 3.1). In particular, in agronomy, animal science, agricultural economics, and agricultural engineering, scientists typically have farm backgrounds. In contrast, only 14% of foresters come from farm backgrounds, while 32.5% report having fathers who were owners or executives of business enterprises. Food scientists, entomologists, and environmental scientists are more likely to come from blue-collar backgrounds than from farm backgrounds.

Of equal importance is the trend over time. How have the family origins of scientists and, particularly, agricultural scientists changed since the early twentieth century? From the data collected by Harmon (1965), it is clear that each new generation of scientists was drawn from a broader social base than its predecessor. As education became more accessible, more members of lower socioeconomic groups received doctorates. The professional portion of the population remained the predominant source of neophyte scientists, but declined relatively as the lower-income groups achieved more education. This was the case for the agricultural scientists in Harmon's sample, as well as the general trend for all scientists. For example, among agricultural scientists receiving the Ph.D. between 1935 and 1940, 12.6% indicated that their fathers were laborers, while 14.9% of the 1945–1950 cohort and 18.0% of the 1955–1960 cohort reported this family origin. Equally interesting is the continuing importance of farm backgrounds for agricultural scientists over this entire period. Harmon (1965) reported that, among the Ph.D. agricultural scientists graduating in 1935–1940, 49.7% indicated that their fathers were farmers or farm managers. That percentage was still 44.7% for agricultural scientists graduating in 1955–1960. For all fields of science, however, there has been a decline in the percentage of scientists from a farm background (1935–1940 cohort, 19.5%; 1945–1950, 15.9%; 1955–1960, 12.7%). This decline is even more apparent in the biological sciences and social sciences, where 21.3% and 22.1%, respectively, of the 1935–1940 cohort were from farm backgrounds. The 1955–1960 cohort in these two groups of sciences reported that only 11.9% and 12.2% of their fathers were farmers.

Our 1979 survey of agricultural scientists reveals similar trends. First, the broadening social base for agricultural scientists has continued for the youngest scientists. Fifteen percent of the scientists born between 1921 and 1930 indicate that their fathers were blue-collar workers, while 21.2% of the birth cohort 1931–1940 and 24.6% of the 1941–1950 group report this family background (Table 3.2). Second, a farm background remains nearly as important for the most

TABLE 3.1
Father's Occupation of Agricultural Scientists by Discipline[a]

Discipline	Farm Background	Professionals	Owners/Executives	White Collar	Blue Collar
Agricultural Economics (113)[b]	54	8	18	10	10
Agricultural Engineering (84)	50	4	18	9	17
Agronomy (279)	54	5	18	5	15
Animal Science (130)	54	9	12	8	15
Basic Sciences (190)	26	16	21	12	25
Entomology (95)	23	8	21	12	34
Environmental Sciences (54)	17	11	19	24	30
Food Science (49)	22	16	22	6	26
Forestry (77)	14	14	32	16	22
Horticulture (91)	31	10	24	11	22
Nutrition (41)	32	15	29	7	12
Plant Pathology (92)	28	16	22	10	17
Social Sciences (47)	28	11	23	13	23
Total (1,397)	38	10	20	10	20

[a]Percentages do not always add to 100 because some of the fathers were employed in occupations not falling into these categories.

[b]Numbers in parentheses represent the respondents in each discipline.

TABLE 3.2
Father's Occupation of Agricultural Scientists by Age Cohort[a]

| Year of Birth (N) | Father's Occupation (%) | | | | |
	Farm Background	Teacher	Owners/ Executives	White Collar	Blue Collar
1907 - 1920 (178)	43	4	21	5	17
1921 - 1930 (344)	39	7	21	10	15
1931 - 1940 (364)	39	5	18	9	21
1941 - 1950 (256)	35	5	19	8	25

[a]Percentage of each age cohort with fathers from the indicated occupation. Numbers have been rounded to nearest whole percentage.

recent entrants into the agricultural sciences as it was for those entering the system twenty to thirty years ago. Indeed, one-third of this youngest age group report farm origins.

Aggregate values conceal important disciplinary differences in these trends. Among the four disciplines reporting the highest percentage from farm backgrounds (agricultural economics, agricultural engineering, agronomy, and animal sciences), the youngest cohorts (29 to 38 years old) report a slight increase over the older cohorts in the percentage of fathers who were farmers. In contrast, in the basic sciences (biology, biochemistry, chemistry, and genetics), entomology, forestry, and horticulture, far fewer members of the younger cohort are from farm backgrounds. Nevertheless, compared to the general population even the youngest cohort of agricultural scientists is twice as likely to have farm origins. In short, despite the rapid decline of the farm population, a large percentage of agricultural scientists still continue to be drawn from farm and rural backgrounds. However, it seems somewhat unlikely that this trend will continue, given the large influx of urban students into the agricultural colleges (Coleman, 1979) and the small size of the current farm population.

To assess some possible consequences of family origins for research, we examined the relationship between father's occupation and each scientist's criteria for problem choice. For ten of the criteria, there are significant differences between scientists with farm backgrounds and scientists with other family origins (Table 3.3). The utility of the research, client needs and demands, and feedback from extension staff are all more important considerations for those with farm backgrounds (Table 3.3). In contrast, scientists with other family backgrounds are more likely to identify criteria that represent a commitment to scientific

TABLE 3.3
Selected Criteria for Research Problem Choice by Father's
Occupation of Agricultural Scientists[a]

Criterion	Father's Occupation	
	Farm Background (534)[b]	Other (865)[b]
Client needs as assessed by you	5.3[c]	4.6
Creation of new methods, useful materials, and devices	5.2[d]	4.9
Demands raised by clientele	4.5[c]	3.9
Publication probability in		
Bulletins	4.0[c]	3.4
Farm/industry journals	3.2[d]	2.9
Potential marketability of the final product	4.0[c]	3.5
Feedback from extension personnel	3.9[c]	3.2
Enjoy doing this research	5.7[d]	5.9
Scientific curiosity	4.9[c]	5.3
Potential contribution to scientific theory	4.1[c]	4.4

[a]Mean scores based on seven point scale (1 = not important,   7 = very important).

[b]Numbers in parentheses represent the respondents in each category.

[c]Row differences, $p < .001$.

[d]Row differences, $p < .01$.

ideals. Here, the more important criteria are scientific curiosity and the potential contribution to scientific theory.

Several questions arise from these findings. Are the relationships between family origin and criteria for problem choice the result of scientists from farm backgrounds consciously integrating their research with their earlier farm experience? Are these scientists more aware of the needs and experiences of the farm sector and, therefore, more likely to consider these factors important in their research? Will these clientele considerations decline in importance if fewer agricultural scientists are drawn from farm backgrounds?

Education

Next, we examined the extent of formal education of agricultural scientists. According to a Bureau of the Census report (U.S. Department of Commerce [USDC], 1978) of persons in fields of science or engineering in 1976, the percentage of all agricultural scientists in both the public and private sectors with bachelor's, master's, and doctorates as the highest degree earned were 54%, 20%, and 25.8%, respectively. The percentage of agricultural scientists with master's degrees or doctorates is considerably smaller than that in the other sciences. For example, in other fields the percentage of scientists with doctorates as the highest degree held is as follows: biologists, 59.5%; medical scientists, 50.6%; chemists, 42.6%; physicists and astronomers, 64.2%; sociologists and anthropologists, 66.4%.

However, within public sector agricultural research, the level of educational attainment among scientists appears to have increased with each generation. In our survey, more than 92% of currently active public sector agricultural researchers have received the doctoral degree. Among the thirteen agricultural disciplines, only two, forestry (81%) and agricultural engineering (63%), have fewer than 90% with doctorates. Moreover, an additional 16% of the agricultural researchers have had some postdoctoral training. This postdoctoral training, however, varies considerably among the agricultural disciplines. Forty percent of the basic scientists report this training, but less that 5% of the agricultural economists, agricultural engineers, and foresters indicate any postdoctoral experience.

Among these agricultural scientists, most have received their education at land-grant institutions. Over 70% of the scientists received their bachelor's degrees, and nine out of ten received their doctoral degrees, at land-grant institutions. In addition, many agricultural scientists received multiple degrees from the same institution. Indeed, 73% of our sample report that they received at least two higher education degrees at the same institution. Following their formal education, agricultural scientists often are employed by the same institution. Among those in our sample currently at a land-grant institution, 31% received at least one degree from their current employer, and 18% received at least two degrees from that same institution (Table 3.4).

Finally, a small number of universities have been extremely important at all levels of the formal education process. At the bachelor's level, for example, approximately one-third of the scientists report having received their degrees from just a dozen land-grant schools. For the master's degrees, two of every five scientists identify one of the top twelve universities. At the doctoral level, this trend is more pronounced with nearly three of every five agricultural scientists having been educated at the top dozen land-grant universities. In fact, just three schools (the University of Wisconsin, Cornell University,

TABLE 3.4
Percent of Land-Grant Scientists with More Than One Educational or Current Employment Experience at the Same Institution by Discipline[a]

Discipline	No Match	Two Degrees	All Degrees	Experience (%)		
				One Degree and Current Employment	Two Degrees and Current Employment	All Degrees and Current Employment
Agricultural Economics (93)[b]	13	49	14	11	9	3
Agricultural Engineering (50)	10	30	4	20	24	12
Agronomy (236)	12	43	11	12	17	5
Animal Science (98)	13	43	6	13	18	4
Basic Sciences (132)	17	46	12	12	11	2
Entomology (64)	8	36	14	17	19	6
Environmental Sciences (43)	16	51	14	7	12	--
Food Science (40)	10	38	12	15	20	5
Forestry (53)	13	46	11	20	8	4
Horticulture (72)	10	46	8	18	12	4
Nutrition (45)	9	48	9	18	9	7
Plant Pathology (72)	21	40	8	11	14	6
Social Sciences (43)	12	47	14	13	12	2
Total (1,089)	13	42	11	14	14	4

[a] Degree is defined as Bachelor's degree, Master's degree, or Doctorate. Percentages are rounded to nearest whole number.

[b] Numbers in parentheses represent respondents in each discipline.

TABLE 3.5
Percent of Agricultural Scientists Trained at the Twelve
Largest Institutions by Degree Received

	Degree		
Institution	Ph.D. N = 1272	M.A. N = 1167	B.A. N = 1368
University of Wisconsin	8.6% (1)[a]	5.6% (1)	3.3% (4)
Cornell University	5.8 (2)	3.2 (6)	2.3 (9)
Michigan State University	5.3 (3)	2.8 (9)	1.8 (16)
Iowa State University	5.1 (4)	3.5 (4)	2.6 (6)
University of Illinois (Champaign-Urbana)	4.9 (5)	4.7 (2)	3.7 (2)
Purdue University	4.3 (6)	2.9 (7)	1.8 (16)
University of Minnesota	4.3 (7)	3.6 (3)	2.6 (7)
North Carolina State University	3.7 (8)	2.5 (14)	1.8 (16)
University of California (Davis)	3.5 (9)	1.5 (23)	1.0 (31)
Oregon State University	2.9 (10)	1.8 (19)	1.6 (18)
Pennsylvania State University	2.9 (10)	3.5 (4)	3.6 (3)
Texas A & M University	2.9 (10)	2.9 (7)	2.6 (7)
Total	54.2	38.5	28.7

[a]Numbers in parentheses represent the rank of the institution in terms of quantity of degrees awarded.

and Michigan State University) have trained one of every five scientists (Table 3.5).

A small number of institutions also play an important role in doctoral education in other fields of science. Analysis of the educational origins of U.S. doctorates between 1920 and 1974 (Harmon, 1978) revealed a similar trend among both agricultural sciences and other sciences. In fact, these data on institutions providing the doctoral training for agricultural scientists are nearly identical to responses from our sample. It is important to note, however, that in each of the other fields of science, the leading degree-granting institutions are not as dominant a source of training as they are in agriculture (Table 3.6). For example, in the biological sciences, physical sciences, and behavioral sciences, the top five institutions have provided the doctoral training for approximately one of every five scientists. In contrast, according to Harmon's (1978) data, the top five doctoral degree–granting institutions in the agricultural sciences have educated approximately one of every three researchers. Finally, as noted earlier, all of the largest doctorate-granting institutions for the agricultural sciences are land-grant universities. In the other sciences, there is a broader spectrum of institutions producing most of the doctoral degrees. In the biological, physical, and behavioral sciences, for example, there are several private institutions (Stanford University, the University of Chicago, Harvard University, Columbia University) and public non-land-grant institutions (the University of Michigan), as well as land-grant universities, among the leading dozen doctoral degree–granting institutions.

This brief review of the educational background of agricultural scientists suggests a potential insularity for both agricultural education and agricultural research. Few public sector agricultural scientists are educated outside the land-grant institutions. In addition, limited mobility is indicated by the tendency of scientists to be employed by their degree-granting institutions and by the dominance of a small number of institutions in the educational process. One of the possible strengths of this system may be the creation of a large community of scholars with a strong informal network and shared sets of assumptions and educational experiences. Potential negative consequences may be inbred stagnation and a limited capacity to respond to new challenges. As Philip Handler of the National Academy of Sciences noted, "Exchange of ideas, of skills, and of styles of life enriches the opportunities for all by introducing variety and by preventing 'the crust of custom' from becoming too hard and heavy" (National Research Council, Office of Scientific Personnel, 1971:v). For science organizations such as university departments to remain vital, they need to be constantly growing and changing. Diversity of educational background can be an important stimulus for innovation and fresh perspectives. In addition, the movement of knowledgeable individuals from one organization or type of institution to another is, perhaps, the most efficient way to transfer knowledge. Consequently,

TABLE 3.6
Percent of Scientists Receiving Doctorates at Largest Degree Granting Institutions by Field of Science

Institutions	All Fields	Field of Science				
		Agricultural Sciences	Biological Sciences	Physical Sciences	Engineering	Behavioral Sciences
Top 5 Institutions	16.1	34.6[a]	19.7[b]	17.8[c]	24.4[d]	19.2[e]
Top 12 Institutions	32.6	58.9	35.7	34.8	41.7	37.0

Source: Compiled from Harmon, 1978:110-111.

[a]University of Wisconsin, Cornell University, Michigan State University, Iowa State University, University of Minnesota

[b]University of Wisconsin, University of California at Berkeley, Cornell University, University of Minnesota, University of Illinois at Champaign-Urbana

[c]University of Illinois at Champaign-Urbana, University of California at Berkeley, Massachusetts Institute of Technology, University of Wisconsin, Harvard University

[d]Massachusetts Institute of Technology, University of Illinois at Champaign-Urbana, Stanford University, University of California at Berkeley, University of Michigan

[e]Columbia University, Harvard University, University of Chicago, University of California at Berkeley, New York University

it appears desirable to reduce institutional insularity among the agricultural sciences by actively seeking qualified scientists with more diverse backgrounds and experiences.

Employment

Agricultural scientists are employed in numerous sectors of U.S. society. The Bureau of the Census survey of all U.S. scientists employed in 1976 reports that the federal government is the leading employer of agricultural scientists (43.6%), followed by business or industry (26.6%) and four-year educational institutions (21.8%) (U.S. Department of Commerce, Bureau of the Census, 1978). Although there is no clear pattern, the U.S. government is seldom the leading employer or even a particularly significant employer of scientists in other fields. However, these statistics are somewhat misleading since only 25.8% of federally employed agricultural scientists have received doctorates, and less than half have education beyond the bachelor's degree. Many of these scientists are laboratory technicians and research assistants rather than principal investigators or leading scientific researchers.

When scientists who received Ph.D.s between 1960 and 1979 are examined separately, the distribution of employment of the agricultural scientists is quite different from that in the more inclusive group (Harmon, 1978). During this period, half of all doctoral agricultural scientists (49.7%) entered academic employment, while only 10.8% and 12.4%, respectively, entered employment with the U.S. government or business and industry. This emphasis on academic employment is consistent with the employment patterns of other Ph.D. scientists. Fifty-four percent of the medical scientists, 61.9% of the bioscientists, 76.3% of the mathematicians, and 67.1% of the behavioral scientists entered the academic market. Only in chemistry (28.1% academic employment) and in engineering (33.3% academic employment), where the academic employment opportunities have been declining for twenty years, was business a primary employer of Ph.D. scientists (54.4% for chemistry and 46.7% for engineering). In the agricultural sciences and other life sciences, academic employment for new Ph.D.s increased from 1960 to the early 1970s, when it declined, first gradually, then more rapidly. During the past decade, as mounting numbers of new Ph.D.s have nearly saturated the academic market, a transition has begun to occur. With few employment opportunities, Ph.D.s in all sciences are beginning to accept a variety of postdoctoral appointments as interim employment. However, at present, a clear majority of all Ph.D. agricultural scientists are employed in the land-grant system or the USDA's research system.

The distribution of agricultural scientists in our survey of principal investigators in the public sector reflects the emphasis on academic employment. Approximately 82% of the agricultural scientists who receive federal agricultural funds are employed by land-grant institutions; the remaining 18% are employed by federal agencies. In addition, just a dozen land-grant institutions employ a third of all the scientists in our sample. This parallels the 1978 data from the annual inventory of agricultural research, which indicated that scientists at just twelve land-grant state agricultural experiment stations generated 34.7% of all the research projects and contributed over a quarter of all scientist-years (USDA, Science and Education Administration, 1981). Once again, the dominance of a dozen institutions is evident in the employment distribution.

Research Orientation

Agricultural scientific research has from the outset been characterized by a tension between the Cartesian image of the scientist as a neutral observer, recording carefully the pageant of nature, and the Baconian image of the scientist as an intervenor in natural systems, changing the world by actively controlling it (Haberer, 1969; Busch, 1980). This division is most clearly manifested in the agricultural sciences in the tension between basic and applied research. Development is, by contrast, of minor importance for public research.

In the past, the need for basic research was often emphasized to business with frequently exaggerated claims that it led to virtually immediate application. Alternatively, basic research was often disguised as applied work to mollify sources of funds, a phenomenon that still exists (Greenberg, 1966). Industry representatives and consultants today find experiment station research to be overly scientific (Cox, 1976; Millar, 1969) but a recent National Research Council report found that there was "grossly inadequate support given to the basic sciences . . ." (National Research Council, 1972:12). The intensity of the debate suggests that these negotiations are likely to continue far into the future. As Richard Levins (1973) remarked, there is a tendency to dichotomize basic and applied research, even to the extent of separating organisms into those suitable for basic research (e.g., *Escherichia coli*) and those suitable for applied research (e.g., wheat rust)!

Scientists often divide their research time among basic research, applied research, and development. The particular orientation they prefer may have major consequences for the selection of research problems and methods, when and where the research is conducted, and, ultimately, the results and products of the work. In addition, as Pelz and Andrews pointed out in their extensive studies on productive research climates, the nature of work may be very important in research effectiveness. They noted "that effective scientists did not limit their activities either to the world of application or to the world of 'pure science' but maintained an interest in both; their work was diversified" (1966:7).

The terms "basic," "applied," and "development," however, are rather vague ones. Indeed, what is characterized as "basic" in one context may be considered "applied" in another (Mulkay, 1977). For example, in applying to the National Science Foundation for a grant, a scientist may emphasize the basic or fundamental character of the research at hand; in describing the same research to a state legislator, the same scientist may well stress its applications. This is not hypocrisy on the part of the researcher; it is simply that different aspects of the same research are stressed in different contexts.

A number of scientists commented on the difficulty of distinguishing among these terms. For example:

> Basic, applied, and developmental phases of research are not unrelated. They are often part of a single research project and a single scientist does all these in fulfilling his mission. On the other hand, some scientists never do basic research but discover truths just the same. Thus, when you imply there is a separation and a balance in these kinds of research, it has to be in concept only.

> I do not agree with the concept of distinguishing between basic and applied research because most of the so-called basic research has applied or practical implications and goals with just a little imagination. Much applied research can be basic.

> I feel the distinction between basic and applied research is a false one, but that development is indeed a somewhat separate enterprise; R&D covers the terrain.

On the other hand, several scientists felt that, in practice, the distinction between basic and applied research often was defined *too* rigorously:

> You should be aware that research tends to the "polar": either it is highly methodological and intended for the use of fellow researchers *or* it is localized and client responsive (emphasis in original).

In order to reduce—though not eliminate—the vagueness of these terms and to assess research orientations, scientists in our study were provided with brief definitions adapted from those used by the National Science Foundation (NSF) (cf. Kidd, 1959):

Basic research: stresses that it is directed toward increases of knowledge in science, with " . . . the primary aim of the investigator . . . a fuller knowledge or understanding of the subject under study, rather than a practical application thereof."

Applied research: is directed toward practical application of knowledge. It covers " . . . research projects which represent investigations directed to discovery of new scientific knowledge and which have specific commercial objectives with respect to either products or processes."

Development: development may be summarized as " . . . the systematic use of scientific knowledge directed toward the production of useful materials, devices, systems or methods, including design and development of prototypes and processes."

Then, using these definitions, scientists specified the percentage of their research time over the last five years spent on basic research, applied research, and development, respectively.

Scientists, on the average, classify 30% of their research as basic, 55% as applied, and 13% as development. However, the totals conceal substantial differences among disciplines (Table 3.7). Not surprisingly, those in the basic sciences (biochemistry, biology, chemistry, genetics) report spending more than half their time engaged in basic research and correspondingly smaller percentages of their time in applied research and development activities. In contrast, in agricultural engineering and agricultural economics, only 15% of research time is considered basic. Agricultural economists and horticulturists report spending the largest percentages of their time in applied research. No group of scientists indicated that they spend more than a third of their time in development activities. Indeed, only agricultural engineers characterize as much as 29% of their work as development.

We wondered whether scientists might exaggerate the percentage of their research time spent in basic research, inasmuch as basic research is often seen as more prestigious. Hence, we also asked them

TABLE 3.7
Reported Research Orientations of Agricultural Scientists by Discipline, Actual and Ideal[a]

Discipline	Research Orientation (%)					
	Basic Research		Applied Research		Development	
	Actual	Ideal	Actual	Ideal	Actual	Ideal
Agricultural Economics	15[w]	22[w]	72[z]	66[z]	12[xy]	12[xy]
Agricultural Engineering	15[w]	21[v]	55[wx]	52[vwxy]	30[z]	27[z]
Agronomy	28[x]	35[x]	61[wxyz]	53[vwxy]	11[xy]	11[xy]
Animal Science	28[x]	37[x]	60[wxy]	52[vwxy]	10[xy]	10[xy]
Basic Sciences	58[z]	60[z]	35[u]	34[t]	7[x]	7[x]
Entomology	31[x]	39[xy]	54[vwx]	47[uvw]	14[xy]	14[y]
Environmental Sciences	27[x]	33[wx]	59[wxy]	55[wxyz]	14[xy]	13[xy]
Food Science	29[x]	38[xy]	56[wx]	48[wx]	16[y]	15[y]
Forestry	24[wx]	30[vwx]	65[xyz]	58[xyz]	11[xy]	12[xy]
Horticulture	14[w]	23[w]	71[yz]	63[yz]	14[xy]	14[y]
Nutrition	46[y]	49[yz]	42[uv]	37[tu]	13[xy]	12[xy]
Plant Pathology	32[x]	39[xy]	56[wx]	50[vwx]	11[xy]	10[xy]
Social Sciences	25[wx]	32[wx]	51[vw]	43[tuv]	25[z]	23[z]

[a] Numbers may not add to 100 due to rounding. For all column differences, $p < .001$.

[t-z] Percentages within a column with different letters are significantly different, $p < .01$.

to characterize the research time utilization of their respective departments. We observed no systematic difference between the self-reports and the departmental figures (Table 3.8).

Scientists were also asked to note the ideal division of their research time (Table 3.7). Here, results across disciplines are markedly similar: Regardless of discipline, scientists wish to devote more time to basic research. Moreover, with few exceptions, they wish to accomplish this by reducing the time spent on applied research, and propose little change in the time devoted to development. An almost identical pattern was found with respect to departments. These findings corroborate those reported for another group of "applied" scientists by Volmer (1972). Many scientists in our study feel quite strongly about the need for more basic research. A toxicologist noted,

> We must continue to encourage foundation studies in science to be able to produce potential answers to new problems of tomorrow—if not, we will continue to lose the world leadership in science, technology, and hence productivity and social status as we know it today.

An agronomist volunteered:

> In my opinion agricultural productivity needs 1) a backlog of basic information, 2) a core of good scientists working between the basic and applied areas, 3) work on practical problems, [e.g.,] quality, environmental quality, marketing etc. The first two areas deserve a minor share of the resources, but both are severely under-funded, at the present time. I also think their role is misunderstood and being poorly communicated. The "image" of basic science is very poor.

Two plant pathologists shared these same concerns:

> I'm interested in what might be eventual or long-run benefits from accumulation of basic knowledge—but in a college of agriculture there is always the pressure to do work that may have some practical application in the relatively immediate future. I've had to balance this with the [basic] work, or do less than I desired, because funding wasn't [available] in the areas in which I was most interested.

> Basic agricultural research is poorly supported in the Universities. The USDA Competitive Grants Program is too small and its funding seems uncertain. In my field (plant virology) we have lost the lead to large labs in France and England.

Furthermore, scientists do not complain that public agricultural research is too basic in its orientation. This suggests that, with available

TABLE 3.8
Perceived Research Orientations of Agricultural Science Departments by Discipline, Actual and Ideal[a]

Discipline	Basic Research Actual	Basic Research Ideal	Applied Research Actual	Applied Research Ideal	Development Actual	Development Ideal
Agricultural Economics	18 u	26 v	66 z	59 z	15 wxy	15 xy
Agricultural Engineering	19 uv	25 v	52 tuvw	49 wxy	28 z	26 z
Agronomy	27 uvwx	37 xy	60 wxyz	51 wxyz	12 uvwx	12 wx
Animal Science	30 wx	37 xy	59 vwxyz	52 xyz	11 uvwx	11 wx
Basic Sciences	48 z	55 z	43 t	37 v	8 uv	8 w
Entomology	37 xy	41 xy	51 tuvw	46 wx	12 uvwx	13 wx
Environmental Sciences	28 vwx	38 xy	64 yz	51 wxyz	8 u	11 wx
Food Science	29 wx	40 xy	54 uvwx	44 vwx	17 xy	16 xy
Forestry	27 uvwx	35 wx	60 wxyz	52 yz	12 uvwx	13 wx
Horticulture	22 uvw	27 vw	63 xyz	56 yz	15 vwxy	16 xy
Nutrition	43 yz	45 y	46 tu	43 vw	11 uvwx	12 wx
Plant Pathology	35 xy	40 xy	55 uvwxy	49 wxy	10 uvw	10 wx
Social Sciences	28 vwx	35 wx	49 tuw	45 wx	21 yz	20 yz

Research Orientation (%)

[a]Numbers may not add to 100 due to rounding. For all column differences, $p < .001$.
t-zPercentages within a column with different letters are significantly different, $p < .01$.

TABLE 3.9
Criteria for Problem Choice by Extent of Research Time Devoted to
Basic Research[a]

Criterion	Percentages of Research Time Devoted to Basic Research	
	Low (0 to 30%)	High (31% to 100%)
Enjoy doing this kind of research	5.7[b]	6.0
Scientific curiosity	4.8[b]	5.7
Contribution to scientific theory	3.7[b]	5.3
Publication probability in professional journals	4.7[b]	5.2
Evaluation of research by scientists in your field	4.2[b]	4.6
Credibility of other investigators doing similar research	3.8[c]	4.2
Importance to society	5.8[b]	5.3
Client needs as assessed by you	5.4[b]	3.9
Potential creation of new methods, useful materials, and devices	5.3[b]	4.6
Demands raised by clientele	4.7[b]	3.2
Priorities of the research organization	4.5[b]	4.0
Marketability of final product	4.3[b]	2.9
Publication probability in experiment station bulletins	4.1[b]	2.8
Currently a "hot" topic	4.0[c]	3.7
Feedback from extension personnel	3.9[b]	2.8
Publication probability in farm or industry journals	3.4[b]	2.3

[a]Mean scores based on seven point scale (1 = not important,   7 = very important).

[b]Row differences, $p < .001$.

[c]Row differences, $p < .01$.

funds and a somewhat changed institutional climate, scientists might substantially increase their focus on basic research.

We suggested earlier that scientists' orientation to basic or applied research or development could affect the way in which they establish research agendas and choose problems. In our survey, it is apparent that research orientation is significantly related to a variety of criteria used in establishing an agenda. Of the 21 criteria for problem choice evaluated by scientists in our sample, 16 are rated significantly differently by scientists with a relatively high percentage of basic research. As Table 3.9 illustrates, scientists devoted to basic research are more likely to consider criteria that represent a commitment to scientific ideals, including scientific curiosity and potential contribution of the research to scientific theory. They also appear to be more discipline-oriented than other scientists. This is reflected by their greater concern for publication probability in professional journals as well as their greater reliance on scientists in their field as a reference group (i. e., greater concern for the "evaluation of research by scientists in your field" and "credibility of other investigators doing similar research").

In contrast, those with a lower percentage of their research time devoted to basic problems are more likely to view client needs and the utility of the research as important for their research program. Table 3.9 indicates nine criteria that are rated significantly higher by this group. On two of the criteria, "client needs as assessed by you" and "demands raised by clientele," the mean differences are greater than 1.5 on a seven point scale. There is, then, a clear and strong relationship between a scientist's research orientation and the criteria considered in developing a research agenda. It is likely as well that this orientation in turn affects the process, products, and dissemination of the research.

4
Scientific Communication

Philip Abelson, the editor of *Science*, recently wrote, "without communication there would be no science." Robert Merton, a well-known sociologist of science, has further noted that "the institutional goal of science is the extension of certified knowledge" (Merton 1973:270). Central and crucial to this process and to the nature and practice of science is a rapid and accurate communication system. Science and its communication seem to be inextricably bound together, so that the production and the dissemination of the results of research go hand in hand. D. W. King et al. also echoed these assumptions in a National Science Foundation report that stated that "communication of scientific and technical information is clearly one of the most important aspects of scientific research, for it is the process that leads to its widespread use and ultimate benefits" (1976:3).

This chapter, therefore, focuses on the communication system in the agricultural sciences, and especially on communication among agricultural scientists actively involved in research. Communication is defined in this instance as the transfer of scientific information from a scientific researcher to his colleagues and from researcher to administrators, educators, clients, and society in general.

Scientific communication in its various forms involves a significant fraction of the scientist's working life. Studies of natural scientists in the United States (Martin and Ackoff, 1963; Menzel, Lieberman, and Dulchin, 1960), for example, revealed that they devoted between a quarter and two-fifths of each working week to formal and informal scientific communication. Formal communication includes various written sources, such as journal articles, books and book chapters, technical reports and bulletins, abstracts, and preprints. It also may include the reading and exchange of papers at professional meetings. Informal channels are usually oral and include face-to-face conversations, telephone exchanges, visits to colleagues' laboratories, and personal correspondence. These two mutually dependent types of communication differ in some important ways. Formal communication is (1) public, with large potential audiences, (2) permanently stored and retrievable, (3) primarily user selected, (4) moderately redundant, (5) relatively dated even when first published, and (6) likely to provide little direct feedback to the author or originator. In contrast, informal communication is (1) usually private, with a restricted audience, (2) typically neither permanently stored nor retrievable, (3) relatively timely or current, and (4) likely to afford considerable feedback to the originator (Garvey and Griffith, 1967). While these two forms of communication may compete for the limited time that any scientist can devote to being a receptor, it should be noted that they are interdependent and complementary sources of information.

Formal Communication

Before examining in detail the formal scientific communication system, it is important to explore why rapid and accurate formal communication is a fundamental requirement of modern science. First, because scientific research is potentially open to a more wasteful duplication of effort than other types of research, knowledge of new research should be diffused as quickly and widely as possible within the scientific community. Second, formal written communication in journals helps keep the scientific enterprise honest. In informal conversations and seminars, scientists are sometimes less than rigorous. However, most are more cautious and thorough in publishing their results, in part because they fear that careless work will be zealously dissected and dismantled by other scientists. Finally, because recognition for a new scientific discovery generally is bestowed upon the person who first announces it publicly, scientists require suitable, trustworthy methods for communicating the results of research, thereby establishing their claims to priority (Mulkay, 1977). Verbal or informal communication is rarely adequate for this purpose; publication, particularly in refereed journals, provides a permanent record coupled with wide dissemination.

Underlying these important functions of formal communication is a combination of psychological and sociological factors that further motivate scientists to participate in the communication process. They may range from the aesthetic pleasure of making new discoveries to more mundane requirements such as the individual's need to gain visibility for his or her work to enhance future professional and economic status. Coupled with this are the complex considerations that link an institution's reputation and pride to its scientists' production of large numbers of ostensibly high-quality published works.

The motives for and expectations about participation in the formal communication system were expressed well by Wooster:

> The young scientist learns, as part of the formal code of behavior of the scientist, that publication of the results of his research in a standard, authorized, refereed scientific journal is not merely right and proper, but a high duty and behaviour expected by his peers and employers. He learns informally that promotion comes about through visibility and that, at least up to a certain critical point in his career, visibility comes about through publication. He learns that there are 'good' journals, and others not so good, but that every manuscript can eventually find a home somewhere and that, for all the platitudes about refraining from unnecessary publication, this must apply to someone else—it is better to publish something in anything, even if only a government report, than not to publish at all (quoted in Meadows, 1974:61).

Given these important functions and motivations for the formal communication system, its enormous growth in the twentieth century is not surprising. This growth has paralleled the rapid increase in the number of scientists noted in Chapter 3. Although statistics in the field of scientific documentation are difficult to compile and often inexact, they provide an approximation of this growth. It is estimated that the annual worldwide production of scientific and technical papers is on the order of 1–2 million papers. In the early 1960s, the total world scientific literature that had been generated since the beginning of modern science contained approximately 12–15 million papers (Meadows, 1974).

The number of scientific and technical journals in the world is estimated to be between 40,000 and 50,000 (King et al., 1976). Despite the rapid increase in the number of papers, however, the annual growth in the numbers of journals has been estimated at only 5% (Meadows, 1974). A main factor, then, in the growth of scientific literature has been the increasing bulk of journals already in existence.

Equally impressive is the rate of consumption of scientific information, which seems to have grown at roughly the same rate as its publication. For example, the circulation of scientific journals in the United States has been doubling roughly every ten years (Meadows, 1974). Abelson (1980) observed that a hundred years ago it would have been possible for a scientist to scan and even read the world's scientific literature, as well as to be acquainted with all of the active research in the United States. Today, a scientist can only be aware of a small fraction of the published work and the ongoing research. Most scientists abandon attempts to be broadly informed and, instead, concentrate on being knowledgeable about a highly specialized field. Later in this book we examine evidence for this pattern among U.S. agricultural scientists.

The United States has been the leader in the production and use of this enormous body of literature. During the first half of the twentieth century, the United States experienced rapid growth in the number of scientists, periodicals and articles. During the last two decades, the growth rate of various U.S. science indicators has leveled off, but the volume and use of the literature remain sizeable. In 1960, the number of scientific and technical journals was 6,300. By 1974, that number had increased to 8,500 (roughly a 35% increase). Although many of these periodicals were trade journals and nonscholarly periodicals, when only scholarly journals were considered, the growth rate over the period remained approximately 30% (Fry and White, 1975; King et al., 1976). This growth rate almost perfectly paralleled the growth in numbers of scientists (correlation coefficient = 0.99). In addition, during this period, the use or consumption of journals also increased. The average number of journal subscriptions increased steadily, from 3,900 U.S. subscribers per journal in 1960 to 6,000 subscribers in 1974 (a 54% increase). The number of journal subscriptions per scientist has also increased slightly, from 3.0 in 1960 to 3.6 in 1974 (King et al., 1976).

Similarly, journal publications by U.S. scientists have grown substantially during this century. By 1960, the annual production of journal articles was 106,000, and by 1974 it had increased 42% to 151,000 (King et al., 1976). Most recently, the actual number of articles published by U.S. authors each year has declined slightly (1973–1979). In addition, from 1960 to 1979, the per scientist prc duction of journal articles decreased slightly. The number of scholarly articles written per scientist fell from 0.091 in 1960 to 0.076 in 1974. However, of the articles published in over 2,100 highly cited and influential journals, the U.S. proportion remained at approximately 37% per annum between 1973 and 1979. Within individual scientific fields, the U.S. proportion of the world's articles in 1979 ranged from 21% for chemistry to 45% for earth and space science. In the field of biology, which includes agriculture and food science, the U.S. proportion was 43% of the world scientific literature (National Science Board, 1981).

Publication of scientific and technical books in the U.S. has also exhibited dramatic increases in the last several decades. From an annual rate of 3,379 titles in 1960, the number of scientific and technical books published per year had increased to 14,442 in 1974 (a 327% increase). In addition, the numbers of conference proceedings, doctoral dissertations, and reports (as measured by sales at the National Technical Information Service [NTIS] and Government Printing Office [USGPO]) have all increased substantially over the past two decades (King et al., 1976).

Within this enormous body of scientific literature, the agricultural sciences and related disciplines represent a relatively small area. Nevertheless, the journals in this field are also increasing in size and number. A fairly recent estimate suggests that some 250,000 papers are published in these journals each year worldwide—approximately 1,000 papers each working day (Tidbury, 1974). In Fry and White's (1975) more selective list of core scholarly journals published in the United States, the number of publications in the life sciences (including agricultural sciences, biological sciences, and medicine) is still substantial. There were 650 scholarly journals in 1974 (an 11% increase from 1960), while journal articles numbered 72,857 (also up 11% from 1960) (King et al., 1976). Interestingly, the life sciences, in comparison with the physical sciences, mathematics, computer sciences, and engineering, have experienced the slowest rate of increase in journal articles during the last two decades. Only the social sciences have shown a slower rate of annual increase in journal article production (King et al., 1976).

With this broad comparative data in mind, we now turn to a careful examination of the role of the formal communication system in current public sector agricultural research. Scientists in our survey were asked to indicate the number of journal articles, books, book chapters, abstracts, bulletins, and reports they had authored and coauthored over the last five years, the journals where their articles were published, and the journals to which they subscribed. In addition, they evaluated the importance of various criteria, including the quantity and quality of publications, for promotion and tenure.

Characteristically, journals are organized along disciplinary lines. In fact, many if not most major scientific journals are published and edited by disciplinary associations. However, in agriculture, the journal publication system is paralleled by a system for publishing in-house reports, which are usually given the title "bulletins." As one might suspect, given the relatively late development of disciplinary societies in agriculture, the use of bulletins to disseminate agricultural scientific information preceded the development of agricultural scientific journals. In fact, one of the earliest U.S. journals—the now defunct *Journal of Agricultural Research*—was itself a USDA project.

The trends in the publication of bulletins and journal articles are apparent from Figure 4.1. Although available data pertain only to experiment stations, the trend probably applies to federal facilities as

78

FIGURE 4.1
Scientific Publications at Experiment Stations, 1896–1975.

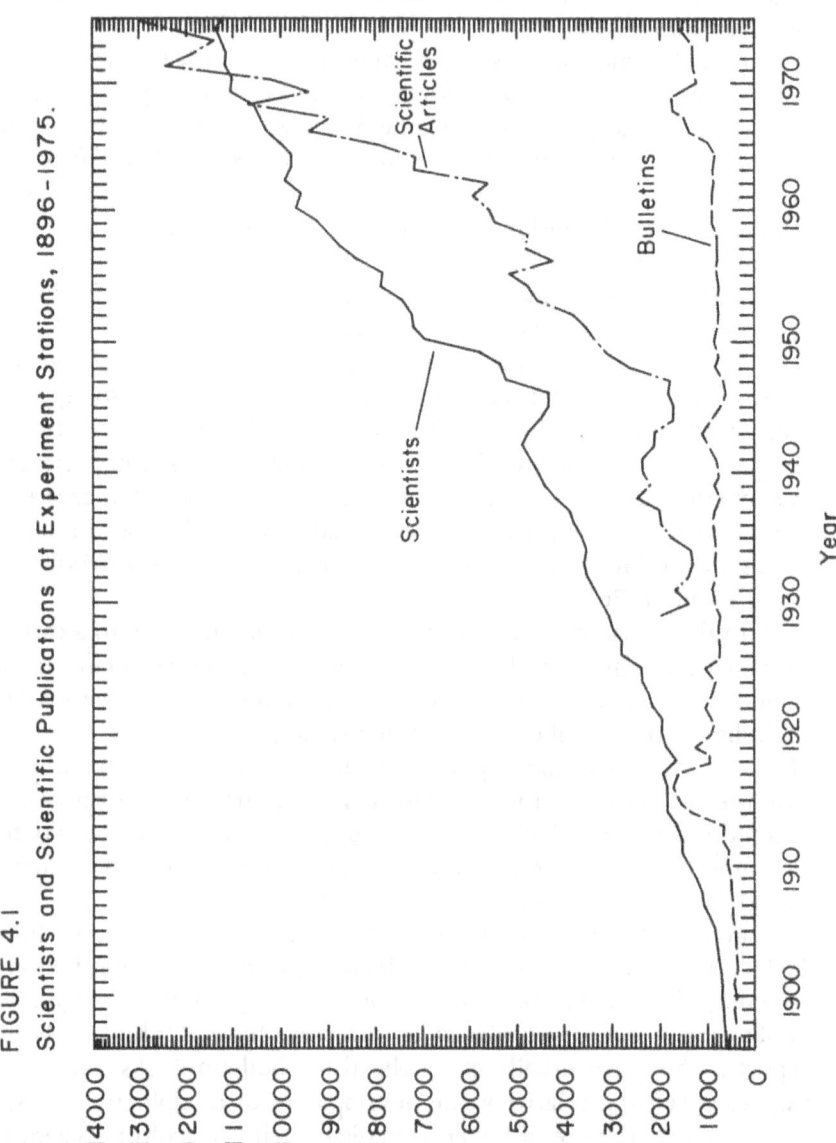

well. According to a contemporary observer, R. Pearl (1915), who believed that increased journal publication was essential to effective research communication, few journal articles were published by agricultural scientists before the First World War. During the first period for which complete documentation is available, 1925–1975, this pattern changed dramatically; the number of articles in scientific journals increased nearly ninefold. At the same time, the number of researchers increased nearly fivefold (USDA, 1925–1975). However, the number of bulletins increased less than twofold over this period (Lacy and Busch, 1982b). One should also note that, while bulletins have been published at the rate of about one for every ten scientists for some twenty-five years, the average production of journal articles has increased from about 0.6 to over 1.0 article per scientist per year since the early 1960s. These figures for SAES contrast with the national data available, which suggest a decrease in journal articles per scientist per year during the same period (King et al., 1976). Unfortunately, changes in the data collection system make it impossible to obtain more recent data. However, all indications are that the trend continues unabated. Several conclusions may be drawn from the data.

First, clearly, the importance of bulletins relative to journals as a publication medium has declined. This has certainly been due, in part, to the rising costs of publishing bulletins. As costs have increased, research administrators have discouraged bulletins and partly under-written journal publication through institutionalization of page charges. In addition, there is some evidence that "the scientist who invents for the market rather than the cathedral [of pure science] loses his membership in the social system of pure science" (Shepard, 1957:119). The fact that journal articles are written for one's peers and not for laymen confers upon them a cachet of quality (Ravetz, 1971) and a certification of the author's membership in the scientific community unattainable through the publication of bulletins. As Ronald Havelock succinctly put it, "In the scientific community the coin of the realm is *recognition* from colleagues" (1971:3–14, emphasis in original).

Second, the increase in per-scientist output of journal articles since the mid-1960s reflects the increased emphasis on publications as a measure of scientific productivity as well as the increased professional specialization of the various agricultural sciences. Third, insofar as journals are the property of disciplinary associations rather than research organizations, this change in the mode of publication reflects a shift toward the disciplinary associations in the power (1) to validate and certify research findings, and (2) less directly, to reward scientists.

Our data on agricultural scientists show that bulletins continued to be of negligible significance, while the importance of journals as

TABLE 4.1
Five Year Scientific Publication Rates Per Scientist
by Type, 1974-1979[a]

	Authorship	
Scientific Literature	Authored	Co-Authored
Journal Articles	4.54	6.64
Books	.09	.11
Book Chapters	.62	.44
Abstracts	2.71	3.57
Bulletins	.96	.99
Reports	4.05	2.59
Other	1.59	.86

[a]Rates based on scientists' self-report of authored
and co-authored publications during the last five years.
For column differences, $p < .001$.

a means of communication continues to increase (Table 4.1). Only 30% of the scientists in our study indicated that they had published a bulletin during the last five years, while more than 80% reported publishing a journal article. If we assume that on the average all journal articles that are coauthored are two-author papers with both members in the land-grant/USDA community, then we can estimate current annual per-scientist publication output as follows:

$$\frac{\text{no. authored articles } + \frac{1}{2} \text{ no. coauthored articles}}{\text{number of scientists}}$$
$$= \text{articles per scientist}$$

Using this procedure, we estimate the current five-year production of journal articles to be 7.9 per scientist, or 1.6 annually. These per-scientist publication figures also illustrate the important role of abstracts (frequently the precursors of journal articles) for the dissemination of research results (Table 4.1). These publication trends, however, are different for different disciplines. The mean number of journal articles, abstracts, and bulletins, authored and coauthored, by field for a five-year period ending in 1979 is summarized in Table 4.2. All fields except agricultural economics conform to the general pattern of publishing more journal articles than bulletins. One possible explanation for this exception is that much agricultural economic analysis is regional or local and, therefore, unpublishable in scientific journals.

TABLE 4.2
Five Year Publication Rates by Type of Publication and by Discipline, 1974-1979 [a]

			Publication			
Discipline	Authored Journal Articles	Co-Authored Journal Articles	Authored Abstracts	Co-Authored Abstracts	Authored Bulletins	Co-Authored Bulletins
Agricultural Economics	1.5 y	2.6 v	0.5 v	0.4 u	2.4 z	3.0 z
Agricultural Engineering	4.6 yz	5.1 wwx	1.5 wxy	0.9 uv	0.8 y	0.4 y
Agronomy	4.3 yz	6.2 wxv	3.6 wxyz	4.4 wxy	0.8 y	1.2 y
Animal Science	5.2 z	7.6 xyz	4.9 z	8.3 z	1.3 yz	1.1 y
Basic Sciences	5.8 z	10.0 z	3.8 xyz	5.3 xyz	0.3 y	0.3 y
Entomology	5.9 z	8.8 yz	1.2 wwx	1.1 uv	1.5 yz	0.6 y
Environmental Sciences	4.0 yz	5.1 wwx	1.1 ww	1.5 uww	0.6 y	0.5 y
Food Science	5.6 z	10.2 z	1.9 wxv	6.2 xyz	0.6 y	0.5 y
Forestry	2.9 yz	4.3 ww	2.0 wxv	1.0 uw	1.0 yz	1.2 y
Horticulture	3.8 yz	5.0 wwx	2.7 wxyz	2.0 uww	0.8 y	1.1 y
Nutrition	4.4 yz	6.1 wxy	4.0 yz	7.4 yz	0.6 y	0.5 y
Plant Pathology	5.2 z	7.1 wxyz	3.2 wxyz	4.0 wwx	0.8 y	0.9 y
Social Sciences	2.7 yz	2.6 v	0.5 v	0.1 u	1.1 yz	1.2 y

[a] For all column differences, p < .001.

u-z Means within a column with different letters are significantly different, p < .01.

All respondents except those in the social sciences report publishing more coauthored than single-authored articles. Moreover, the fields with the highest rates of journal publication—the basic sciences and food science—also have the strongest tendency toward multiple authorship. This reflects the increased scale and corporate character of much agricultural research.

The various disciplines differ markedly in their use of abstracts as a communications medium. Abstracts are insignificant in economics and the social sciences, but common in animal science and nutrition. Other types of publications are common as well, but rates do not significantly differ by discipline. For example, during the five-year period covered by our data, the average scientist authored or coauthored 0.2 books, 1.1 book chapters, and 6.6 reports (Table 4.1).

Although scientists depend heavily on professional journals as an outlet for their results, they appear to restrict their publications to a small number of journals within their own disciplines. Despite authoring or coauthoring an average of approximately eleven journal articles during the last five years, three-fifths of the scientists report having published in three or fewer journals during that period. Indeed, only 8% of the scientists published in more than six journals.

Besides being an important outlet for research results, the formal communication system is also seen as central to the research process itself. Scientists identify U.S. journals, particularly those in their fields or disciplines, as the most important published resource for their research. Over 60% of the scientists indicate that domestic journals in their fields are very important resources for research (Table 4.3). Second in importance are domestic journals in related fields, followed by books and monographs and foreign journals in their fields or disciplines. The least important published resources are research bulletins and foreign journals in related fields. The dominance of domestic journals and the relative neglect of foreign journals, particularly those outside one's field, are consistent with previously reported findings among other scientists (Gerstl and Hutton, 1966; Meadows, 1974). For example, Shaw (1956), in a survey of U.S. scientists and engineers, reported that 70% of all their reading was in journals. In addition, Meadows (1974) noted that neither British nor U.S. scientists made much use of foreign language publications, for they averaged less than 5% of their total reading time on such publications.

Among disciplines, there are, again, significant differences in scientists' assessments of published resources as important for their research. Domestic and foreign journals in one's field, as well as in related fields, are ranked significantly higher by basic scientists, food

TABLE 4.3
Perceived Importance of Published Resources for Research[a]

Resource	Mean Score
Domestic journals in own field	6.4
Domestic journals in related fields	5.3
Books and monographs	4.9
Foreign journals in own field	4.7
Research bulletins	4.4
Foreign journals in related fields	3.5

[a]Mean scores based on seven point scale (1 = not important, 7 = very important). For column differences, p < .001.

scientists, and plant pathologists than by agricultural economists and agricultural engineers. In contrast, agricultural economists, social scientists, and foresters are more likely to identify books and monographs and research bulletins as important resources.

This emphasis on disciplinary journals is also reflected in the journal subscription rates. Scientists reported subscribing to an average of nearly four journals ($\bar{x}$ = 3.8). However, over half subscribe to three or fewer journals. Most individual subscriptions are concentrated in a couple of fields, with the mean number of fields in which scientists have at least one subscription equal to 2.4.

A further indication of the importance of journals in the research process is the scientists' ratings of criteria for their choice of research problems over the last five years. From a list of 21 criteria, "publication probability in professional journals" is the sixth most important criterion for research problem choice (Busch, Lacy, and Sachs, forthcoming). Over 40% of the scientists indicate that "publication probability in professional journals" is a very important criterion (6 or 7 on the 7 point scale). On the other hand, "publication probability in bulletins and reports" is nineteenth in importance, and "publication probability in farm and/or industry journals" is last as a criterion for problem choice (Table 2.1).

Finally, the formal communication system is viewed as playing a major role in the criteria for promotion and reappointment. Scientists were asked to assess the importance of various factors in their reappointment, promotion, and tenure in their department or institute. Publication of many scholarly papers is seen as the most important criterion, and publication of high quality scholarly papers as second. Publication of in-house papers and reports is a distant fifth. In indicating the factors that should ideally be employed in these decisions, scientists show a preference for the quality of the publications to be the major factor and a desire to deemphasize quantity. However, both publication criteria are among the top three factors under the ideal policy. (For further discussion of the important role of journals in agricultural science see Lacy and Busch, 1982b.)

Given this central role of formal communication in research, it is important to explore the relationship between: (1) preferences for various modes of formal communication as research inputs and outputs, and (2) the process of establishing research agendas. We began this chapter by pointing out that the process and practice of science and its communication are closely related. This becomes apparent in the correlations between the importance placed on various published resources for one's research, and the criteria for a scientist's choice of research. Scientists who rate domestic and foreign journals in both their own fields and related fields as relatively more important research resources differ significantly from other scientists in the criteria they identify for problem choice. These scientists are more likely to identify the following discipline-oriented criteria as important for their research: its potential contribution to scientific theory, scientific curiosity, their enjoyment in doing this kind of research, the availability of research facilities, the evaluation of scientists in their field, and publication probability in professional journals (Table 4.4). These same scientists are less likely to view marketability of the final product, and demands, needs, and feedback from clientele, as important to their choice of research. In contrast, those scientists who view research bulletins as relatively important to their work are more likely to be concerned with clients' needs and demands, feedback from extension, and marketable final products in selecting research topics. They also deemphasize scientific disciplinary criteria such as one's contribution to scientific theory and scientific curiosity. The same association exists between the diversity of journals to which one subscribes and one's criteria for problem choice. The wider the range of fields or disciplines covered by a scientist's journal subscription list, the more scientifically oriented his or her criteria for choosing a research agenda (Table 4.5). Finally, the diversity or breadth of a scientist's use of resources may have further implications that were not examined in this study. For example, one study of creativity among research chemists at a

TABLE 4.4
Correlations Between Importance of Published Resources and Selected Criteria for Problem Choice[a]

| Criteria | Published Resources | | | | |
| | Domestic Journals | | Foreign Journals | | |
	Own Field	Related Fields	Own Field	Related Fields	Research Bulletins
Contribution to scientific theory	.17	.25	.28	.32	-.24
Marketability of product	-.12	-.13	-.14	-.09	.25
Publication in					
Professional journals	.22	.22	.15	.16	-.12
Bulletins	-.12	-.17	-.22	-.19	.49
Availability of facilities	.17	.13	.19	.15	.00
Evaluation by scientists in your field	.16	.15	.11	.12	-.01
Enjoy doing this kind of research	.08	.15	.12	.15	-.08
Scientific curiosity	.12	.20	.16	.19	-.14
Demands of clientele	-.06	-.11	-.16	-.15	.34
Feedback from extension	-.04	-.11	-.14	-.14	.33
Client needs as assessed by you	-.07	-.14	-.19	-.21	.34

[a]Pearsonian correlation coefficients (r). Values range from -1.0 (strong negative relationship) to 1.0 (strong positive relationship). A value of 0.0 is interpreted as the complete absence of any relation between the two variables. As the data is ordinal and the coefficients assume interval data, reported results are conservative. If $r > \pm .08$, $p < .001$; if $r > \pm .06$, $p < .01$.

large U.S. chemical company found that those rated as highly creative typically consulted twice as much literature outside their own field as those classified as having a low level of creativity (Menzel, Lieberman, and Dulchin, 1960).

The relationship between scientific productivity, as measured by publications, and criteria for problem choice parallels the association between resources and criteria. Publication of authored and coauthored journal articles and abstracts is positively related to the importance placed on contributions to scientific theory and scientific curiosity as criteria for one's research choices. Similarly, relative reliance on bulletins as an important form of communication is linked to client and extension-oriented criteria. The relationship between journal publication diversity and criteria for problem choice parallels that between subscription diversity and problem choice (Table 4.5). The greater the number of fields in which a scientist has published, the more likely he or she is to view scientific criteria as important for establishing a research agenda.

In summary, agricultural scientists view formal communication, and

TABLE 4.5
Correlations Between Journal Diversity and Selected Criteria for Problem Choice[a]

	Journal Diversity	
Criterion	Subscriptions[b]	Publications[c]
Contribution to scientific theory	.13	.18
Marketability of product	-.11	-.11
Time to complete research	-.08	-.10
Publication in professional journals	.05	.12
Bulletins	-.16	-.26
Farm journals	-.12	-.19
Currently a "hot" topic	-.12	-.09
Enjoy doing this kind of research	.08	.11
Scientific curiosity	.14	.14
Demands of clientele	-.11	-.17
Feedback from extension	-.13	-.15
Client needs as assessed by you	-.11	-.14
Priorities of the organization	-.10	-.06

[a]Pearsonian correlation coefficients (r). Values range from -1.0 (strong negative relationship) to 1.0 (strong positive relationship). A value of 0.0 is interpreted as the complete absence of any relation between the two variables. As the data is ordinal and the coefficients assume interval data, reported results are conservative. If $r > \pm .05$, $p < .005$.

[b]Journal subscription diversity is measured by the mean number of fields in which scientists have at least one subscription.

[c]Journal publication diversity is measured by the mean number of fields in which scientists have at least one publication.

particularly professional journals, as important resources in their research, major outlets for their findings, a key criterion in their choice of research problems, and an important factor for promotion and reappointment. Modes of formal communication and the processes of scientific research itself are closely related and often inseparable.

Informal Communication

While the formal communication system is highly visible and central to the research process, the informal network is more difficult to observe but equally important. Detractors of informal communication have argued that is is marked by a large random element. Its use is often seen as a sign of weakness and the need for better formal systems (C. Scott, 1962). However, it is a misinterpretation to assume that informal means lacking structure. Although informal systems by definition lack formal structure, within the research system, the informal network may consist of a rich system of interrelations. In addition, an increasing number of social researchers have concluded that people and not formal channels of communication are more effective for transmitting technical information (Allen, 1970; Ganz, 1976; Kelly et al., 1977). Most surveys of scientists indicate that discussions with colleagues rank with journal articles as the major sources of research information (Meadows, 1974). Indeed Garvey and Gottfredson (1977), in a study of two thousand scientists, reported that most scientists (63%) had prepublication knowledge of journal articles and that the most likely source of this information was face-to-face contact. Garvey, Lin, and Tomita (1972) further noted in a study of nine scientific disciplines that scientific information in most journal articles is six months to a year behind the research front and that communication in the informal domain brings about genuine mutual information exchange. Another survey of U.S. biochemists found that younger scientists obtained almost twice as many ideas from discussions as from the literature (Shilling, Bernard, and Tyson, 1964). In Rosenbloom and Wolek's (1970) study, 650 scientists were asked to describe the most recent instance in which they had received information useful in their work from a source other than someone in their immediate circle of colleagues. The three general categories of responses that emerged were: (1) information that had been sought for the specific use for which it was employed (42%), (2) information that had been volunteered by someone else (33%), or (3) information that had been acquired in the course of competence development activities such as reviewing the literature in a field (25%). This demonstrates the immense value of the informal network in transferring information even before the scientist is aware of the need for it.

The relative significance of literature and personal contacts as sources of information varies with the nature of the research as well as the country in which the scientist works. For instance, researchers in the basic sciences obtain relatively more information directly from the literature; applied scientists, and engineers in particular, tend to rely more on personal advice and less on journals (White, 1970; Gerstl and Hutton, 1966). In one study of scientists and engineers, the average number of papers read in detail varied from a low of 1.5 per week for engineers to a high of 7.4 for scientists in basic medical research (Bernal, 1948). Moreover, Pelz and Andrews (1976) found that Ph.D.'s in development laboratories and engineers reported a higher frequency of personal communication about work matters than did Ph.D.'s in research laboratories. In addition, there are some indications that increasing specialization produces a greater reliance on personal contacts for information, since scientists find it more difficult to evaluate the relevance of research outside their discipline to their own work (Meadows, 1974). A rapidly developing discipline or area is likely to be highly competitive and to contain many people who are unknown to each other. This may restrict informal communication. As an example, a study of scientists from a rapidly developing field, biochemistry, and a more established field, zoology, revealed that transfer of information via personal contacts was easier for zoologists (Meadows, 1974). Finally, U.S. scientists place appreciably greater reliance on personal contacts than do scientists in other countries (Paisley, 1966). In part, this is due to the greater opportunity for frequent personal contacts in the United States, which has a larger concentration of scientists.

Informal communication has become such a strong competitor for the limited amount of time that any scientist can devote to receiving information that Abelson observed that scientists should ask themselves "Is anybody reading?" (1980:61). Kelly and Wolek have gone so far as to advise, "When you need to know something, ask somebody. If you don't know who to ask, ask someone who might know somebody. If you don't know someone who knows somebody, and can't wait, avoid, or change your need, then search the literature" (quoted in Kelly et al., 1977:26). Of course, to relegate the formal communication system to this inferior role is an exaggeration. Requests for information may well be met with a reference to the literature where the information could be found. However, the informal system is increasingly regarded as an essential dimension of the scientific process.

To assess the nature and influence of informal communication for agricultural scientists, researchers in our study were asked to report the frequency with which they communicated with a variety of groups of people about their research. They were also asked to indicate the groups of people they saw socially. Finally, they indicated the degree to which eleven different types of participants in the research system influenced their research decisions concerning choice of research problem, methods, theoretical orientation, and indentification of key concepts.

Responses received to the question, "How frequently do you communicate with the following people regarding your research?" are graphically illustrated in Figure 4.2. Scientists report communicating about their research with scientists in their own departments somewhat less than weekly. They report discussing their research with agricultural scientists outside their departments somewhat less than twice a month. Discussions about their research with other (nonagricultural) scientists, administrators, clients, funding agencies, and extension staff apparently take place less than once a month.

The relative frequency of scientists' communication with each group

FIGURE 4.2
Scientists' Frequency of Communication Regarding Research with System Participants.

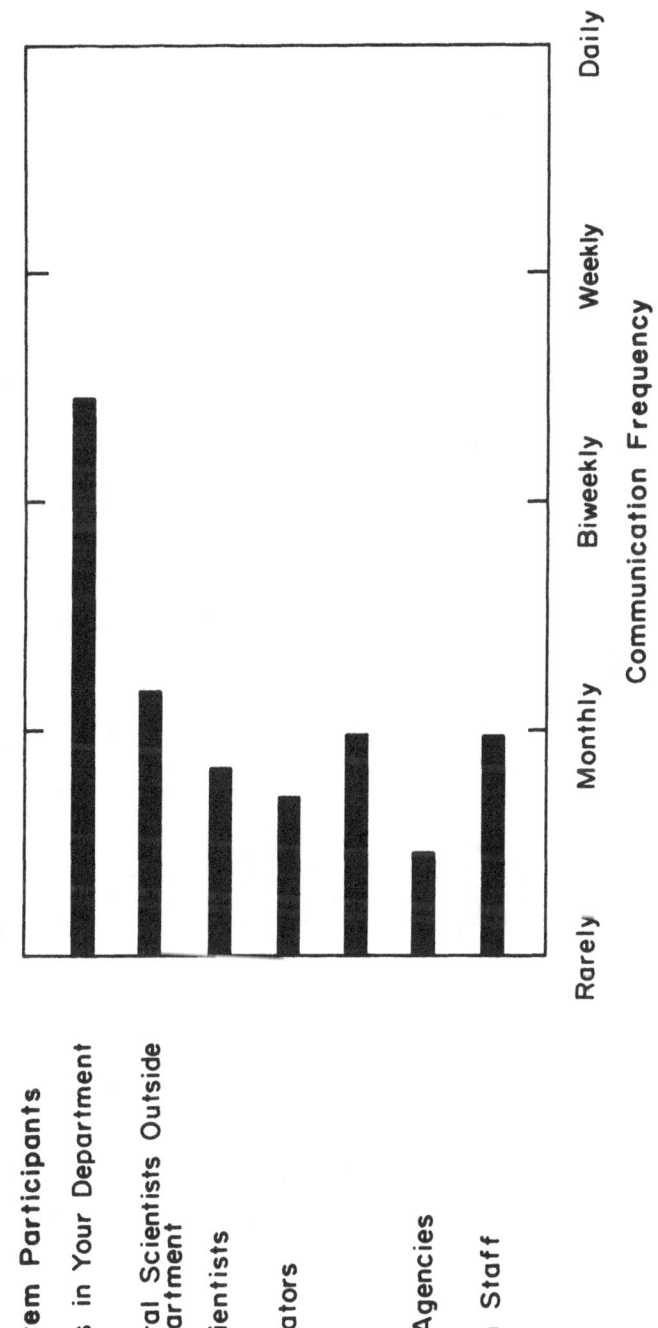

TABLE 4.6
Frequency of Communication Concerning Research with System Participants [a]

Discipline	Participants						
	Scientists in Your Department [c]	Agricultural Scientists Outside Department [c]	Nonagricultural Scientists [b]	Administrators [c]	Clients [b]	Funding Agencies [b]	Extension Staff [b]
Agricultural Economics	3.2 wxyz	1.9 ww	1.4 v	1.6 xy	1.9 xyz	1.4 wxy	1.9 xy
Agricultural Engineering	3.7 z	2.5 z	2.0 wxyz	2.0 z	2.3 z	1.6 yz	2.1 xyz
Agronomy	3.6 xyz	2.4 yz	1.7 vw	1.8 xyz	2.1 z	1.4 wxy	2.3 yz
Animal Science	3.4 wxyz	2.3 vwxyz	1.7 vwxy	1.8 xyz	2.0 yz	1.4 wxy	2.2 yz
Basic Science	3.4 wxyz	1.9 vwx	2.1 yz	1.6 xy	1.5 x	1.4 x	1.4 vw
Entomology	3.3 wxyz	2.3 wxyz	1.8 vwxy	1.5 x	1.9 xyz	1.3 wxy	2.1 xyz
Environmental Science	3.1 wxy	2.0 vwx	2.2 z	1.6 xy	2.1 yz	1.7 z	1.8 vwxy
Food Science	3.3 wxyz	2.0 vwxy	1.6 vw	1.5 x	2.0 yz	1.4 wxy	1.3 v
Forestry	3.6 yz	2.0 vwx	2.0 wxyz	1.9 yz	2.2 z	1.3 wx	1.6 vwx
Horticulture	3.1 wx	2.4 xyz	1.5 v	1.7 xyz	2.3 z	1.3 wx	2.5 z
Nutrition	3.0 w	1.8 v	1.8 vwxy	1.7 xyz	1.6 xy	1.3 wx	1.8 vwxy
Plant Pathology	3.5 wxyz	2.5 z	1.7 wx	1.6 xy	2.0 yz	1.2 w	2.3 z
Social Science	3.4 wxyz	1.8 v	2.1 xyz	1.8 xyz	1.9 xyz	1.5 xyz	1.8 wxy
Overall Mean	3.4	2.2	1.8	1.7	2.0	1.4	2.0

[a] Mean scores based on five point scale (1 = rarely, 2 = monthly, 3 = bi-weekly, 4 = weekly, 5 = daily).

[b] Column differences, p < .001.

[c] Column differences, p < .01.

v-z Means within a column with different letters are significantly different, p < .01.

generally remains similar within each discipline (Table 4.6). For example, scientists in all disciplines are most likely to communicate with colleagues in their own departments (albeit only weekly) and least likely to interact with funding agents. However, the degree of interaction across disciplines does vary. For example, agricultural engineers are more likely than other agricultural scientists to communicate with a variety of participants in the research system. On the other hand, agricultural economists are less likely to communicate with any group of scientists outside their department. Basic scientists and social scientists shared this pattern of relatively infrequent communication with other agricultural scientists, but are more likely to communicate with nonagricultural scientists. Basic scientists are also less likely to communicate with clients and extension agents. In contrast, horticulturists are more likely to commmunicate with these same two groups. Despite these differences, however, the relatively low levels of informal scientific communication, particularly communication with those outside one's department, reflect the extensive disciplinary separation characterizing most scientific institutions.

Several researchers in our study were concerned about the lack of contact or communication among agricultural scientists. One scientist commented that "scientists are now so specialized that they become isolated from many others and tend to interact only with fellow scientists in the same areas of research. Biological scientists appear too far removed from social scientists to cooperate on joint projects." To counter this isolation, he recommended that a person first be educated (presumably this would entail a broad-based understanding of the humanities, social sciences, and natural sciences), "then train the person as a scientist." A plant physiologist recommended that "the USDA take more of an intellectual lead and foster more interaction among basic scientists and between basic and applied scientists."

The infrequency of social contact with fellow scientists within or outside the department is even more pronounced. Scientists in no discipline report even monthly social contact with colleagues outside their department. Social contact with colleagues within one's department is not much greater, with an average frequency of monthly interaction. In terms of frequency of social contact, however, agricultural scientists are not asocial people. They report meeting socially with neighbors and relatives on the average of once or twice a week. The data on social contact suggest that informal social exchanges between scientists rarely occur and are not likely to compensate for relatively infrequent communications with other scientists about one's research.

Relative frequency of communication with colleagues and clients is only one way to assess the importance of informal communication in research. Various people may be influential in shaping one's research without frequent direct contact. Therefore, scientists in this study were asked to indicate the degree of importance of a variety of potentially influential people for their choice of research problem, research methods, key concepts, and theoretical orientation. The possible influential people included one's immediate supervisor, a department colleague, an agricultural scientist at one's institution, a basic scientist at one's institution, a colleague at another institution, a research associate or technician, a graduate student or postdoctoral student, a former professor, the director of one's research facility, a client or potential user, and a research review committee. The degree of influence of these people varies considerably. In many instances, over half of the scientists indicate that a potential source of influence was of no importance to their research decisions. However, a majority of scientists report that some person was at least moderately important in decisions on problems, methods, concepts, and/or theoretical orientation. Indeed, over 90% of the researchers indicate that someone influenced their choice of research problem. In addition, at least one person is identified as having influenced choices of methods for 88% of the scientists, as having influenced identification of key concepts for 80%, and as having influenced theoretical orientation for 70% of the scientists.

Clients or potential users emerge as the most frequently mentioned group that significantly influences scientists' choice of research problem or topic (Table 4.7). This is particularly interesting since the frequency of contact with clients is relatively limited (Table 4.6). However, clients' influence on other research decisions, such as choice of methods, is generally minimal. The immediate supervisor plays a similar role for a number of scientists, with a strong influence on research problem choice but a relatively insignificant influence on other decisions.

One of the most important sources of influence, not only for problem choice but for all research decisions, is colleagues at another institution. Despite the scientists' relatively infrequent communication with these colleagues, a fifth of all agricultural scientists report that colleagues at another institution are among the most important people in influencing choice of problems and research methods (Table 4.7). Given this influence, it is important for future research to explore the means by which effective contacts occur. Though little is known about how these collegial relations develop, it is likely that professional meetings and conferences are instrumental in promoting them. Garvey et al. (1972) noted that the first public announcement of a large portion of current research findings in any discipline is made at a

TABLE 4.7
Persons Influential in Scientists' Research Decisions[a]

Influential Persons	Research Decision (%)			
	Research Problem	Research Method	Key Concepts	Theoretical Orientation
Immediate supervisor	20	4	6	4
Department colleague	18	19	16	10
Agricultural scientist at one's institution	11	11	8	6
Basic scientist at one's institution	3	7	5	5
Colleague at another institution	20	19	18	12
Research associate or technician	4	10	4	2
Graduate or postdoctoral student	11	12	9	5
Former professor	11	9	9	9
Director of one's research facility	12	2	3	3
Client or potential user	36	5	7	3
Research review committee	11	7	5	4

[a]Numbers represent the percentage of scientists who identified particular persons as important or most important for a research decision (5, 6 or 7 on a seven point scale from 1 = the person was slightly important, to 7 = most important).

national meeting. In addition, despite the increasing size of these meetings and their formalization, the participants regard the national meeting as one of the last informal scientific communication media and a crucial source of new information. This is particularly so for the younger scientists. One scientist in our study wrote, "I think that scientific meetings provide excellent communications and motivation." A plant physiologist volunteered that professional society meetings were important in "stimulating areas of new research, new methodology, and updating scientists' awareness of new discoveries. . . . I have found that contacts made at such meetings have had a very significant impact on my research and my colleagues." If these meetings are generally as important for inter-institutional scientific communication as suggested by the literature (Meadows, 1974; Garvey et al., 1972) and the scientists in our sample, then restrictions on conference attendance could seriously disrupt this form of communication. One scientist observed that

the federal government has recently put detrimental constraints on expenditures for travel for its scientists, a self-defeating and counter-productive measure to 'reduce spending.' It would be valuable to document the impact of such travel (in relation to attendance at scientific meetings) on scientific progress. These meetings are an essential mechanism for assuring relevance of scientific research effort to society's goals and needs.

Another important source of influence much closer to home is one's departmental colleagues. Not surprisingly, they play a key role for a number of scientists in research decisions on choice of problem or topic, methods, key concepts, and theoretical orientation (Table 4.7). Approximately three-fifths of all scientists indicate at least some influence by departmental colleagues on research methods and problem choice. Indeed, nearly a fifth of the scientists report that departmental collegues have been most important for these two types of decisions.

A smaller but significant percentage of scientists identify additional important sources of influence. Approximately a tenth of all scientists see a former professor as exerting strong influence on the full range of research decisions. Similarly, agricultural scientists in another department at one's institution strongly influence about a tenth of the scientists in their choices of problem, methods, and key concepts. The director of one's research facility, however, is limited to influencing research problem choice and has essentially no influence in other areas. In contrast, research associates or technicians are an important source of influence on choice of research methods for one in ten scientists but a negligible influence on other decisions. Graduate or postdoctoral students, however, have a wider range of influence for about a tenth of the scientists. Scientists in basic science departments appear to be uniformly noninfluential for nearly all the scientists' research decisions (Table 4.7) Over 55% of all agricultural scientists report that such scientists are of no importance to any type of research decison. Only in decisions on research methods are scientists in basic science departments identified as very influential for as many as one in every 14 scientists (Table 4.7). This strongly implies that the overall interaction between members of basic science and agricultural science departments is weak.

The sources of influence on various research decisions vary across the agricultural disciplines. For example, a strong influence of clients or potential users on research problem choice is most likely to occur among agricultural engineers, foresters, agricultural economists, and horticulturists. Their influence on problem choice among social scientists and basic scientists is negligible. Similarly, one's immediate supervisor is most likely to influence problem choice among agricultural engineers, foresters, and agricultural economists, and least likely to shape research agendas among social scientists and basic scientists. Departmental colleagues appear to be particularly important sources of influence on research methods and identification of key concepts among agricultural economists.

The nature of social influences on research decisions may also vary between university scientists and federal government scientists. For example, Meadows (1974:227) concluded that the "efficiency of government or industrial laboratories depends on the effectiveness of their internal communication channels, whereas academic laboratories depend more on the effectiveness of their communications with other, external groups working in the same discipline." This generalization is reflected in the patterns of communication among agricultural scientists in our study. Governmental scientists are more likely than university researchers to indicate that immediate supervisors influence their problem choice and research methods. In contrast, university scientists view colleagues in other departments and institutions as relatively more important for decisions on research problems and methods.

Despite the infrequency of scientific communication among researchers and the variations in the frequency and patterns of interpersonal communications within disciplines and organizations, informal communication appears to be an integral part of all scientists' research. The remaining question concerns its relationship to the other processes and products of science. Does it matter? Previous studies of informal communication among scientists suggest that it does. For example, Meadows (1974) and Allen (1969) concluded that, in universities, the greater the extent to which members of a department involved themselves in extradepartmental communication, the greater their productivity. Pelz and Andrews (1976), in a study of 1,300 scientists and engineers, concluded that communication with any colleagues on work-related matters was directly related to scientific performance. Performance was measured by published papers, unpublished manuscripts, and senior colleagues' judgments of the scientists' work. Ph.D. scientists in development and in research laboratories and engineers who had relatively frequent contact with colleagues tended to perform at higher levels than those with less frequent contact. While the median frequency of contact among Ph.D.'s in research laboratories was about weekly (comparable to our data in Figure 4.2), scientists who contacted colleagues semiweekly or daily had the best performances. Furthermore, colleague contacts were positively related to performance whether the scientist being assessed initiated the contact or not. Finally, "the relationships between contacts with colleagues and performance were as positive among scientists who did not especially prefer to work with others as they were for scientists who did prefer to work with others" (Pelz and Andrews, 1976:51). Indeed, among Ph.D. scientists in research laboratories, the relationships were somewhat stronger for the relatively asocial scientist. Pelz and Andrews concluded that frequent contact with many colleagues was positively related to performance for all groups of scientists.

Our research permitted an examination of related questions regarding the link between informal communication among agricultural scientists and other scientific processes and products. Earlier in the chapter we observed a relationship between scientists' perceptions of the importance of formal communication and the criteria they use in choosing a research problem. These criteria are also related to the frequency of communication with various participants in the research system (Table 4.8). Scientists who frequently communicate with non-agricultural scientists are more likely to employ scientifically oriented criteria (such as scientific curiosity and the potential contribution to scientific theory), in selecting a research problem. On the other hand, frequent contact with clients and extension staff, and to a lesser extent with administrators, is negatively related to scientific criteria for problem choice. Relatively frequent contact with these latter three groups, as well as contact with agricultural scientists outside one's department, is highly and positively associated with criteria stressing client needs and utilitarian qualities. This is particularly true for scientists who frequently communicated with clients and extension staff (Table 4.8). Finally, for those scientists frequently communicating with nonagricultural scientists, client demands and the instrumental nature of the research being considered are less likely to be important criteria.

Similar relationships exist between the people identified as important for research decisions and the important criteria for one's choice of research problems. Scientists who report that basic scientists, former professors, postdoctoral or graduate students, and colleagues at another institution influence their work are also more likely to consider a topic's potential contribution to scientific theory and their own scientific curiosity as important criteria for selecting their research. In contrast, scientists more influenced by clients or potential users deemphasize these scientific criteria and are more likely to identify client needs, feedback from extension, potential marketability of the final product, and publication probability in farm journals and bulletins as important criteria for problem choice. In addition, when one's immediate supervisor, the director of one's research facility, or a research review

TABLE 4.8
Correlations Between Frequency of Communication with System Participants and Criteria for Problem Choice[a]

	Participants				
Criterion	Agricultural Scientist Outside Department	Non-agricultural Scientists	Administrators	Clients	Extension Staff
Contribution to theory	.05	.18	-.12	-.25	-.23
Marketability of product	.16	-.10	.15	.34	.22
Publication in Farm journals	.17	-.12	.20	.30	.28
Bulletins	.14	-.17	.19	.26	.31
Importance to society	.16	.09	-.10	.20	.17
Scientific curiosity	.17	.17	-.13	-.17	-.14
Demands of clientele	.14	-.12	.19	.45	.33
Feedback from extension	.19	-.13	.16	.32	.55
Client needs as assessed by you	.17	-.14	.15	.44	.35
Priorities of research organization	.13	-.09	.19	.15	.12

[a] Pearsonian correlation coefficients (r). Values range from -1.0 (strong negative relationship) to 1.0 (strong positive relationship). A value of 0.0 is interpreted as the complete absence of any relation between the two variables. As the data is ordinal and the coefficients assume interval data, reported results are conservative. If $r > \pm .10$, then $p < .001$.

committee has a significantly high degree of influence on a scientist's orientation, "priorities of the research organization" is likely to be the most important criterion for problem choice. Other relatively important criteria among scientists influenced by these three sources include client demands, feedback from extension personnel, publication probability in bulletins, and evaluation of the research as a hot topic. At the same time, those more apt to be influenced by immediate supervisors or research review committees are less likely to regard scientific criteria as important for their choice of research.

Not surprisingly, our examination of the relationship between communication frequency and publication productivity reveals several positive associations. Scientists who more frequently communicate with nonagricultural scientists are more likely to publish single-authored and coauthored journal articles and book chapters. In contrast, those who more frequently communicate with clients, extension staff, and agricultural scientists outside their departments are more likely to publish single-authored and coauthored bulletins and reports.

Conclusions

Mutually interdependent formal and informal scientific communication systems are integral to agricultural research. U.S. agricultural scientists identify modes of formal communication (particularly professional journals) as: (1) essential resources and important criteria for their choice of research problems, (2) the key to the scientists' reward structure, and (3) major outlets for research results. Similarly, these same scientists view the numerous individuals in the informal communication network as important sources of influence on a variety of research decisions.

In spite of its perceived importance, scientific communication among agricultural scientists is relatively infrequent, highly specialized, and insular. Despite the enormous growth of published scientific materials during the last few decades, especially in foreign countries, use or consumption as measured by journal subscriptions has remained relatively constant at approximately three domestic, primarily discipline-oriented journals per scientist. The outlets for scientists' work also are dominated by domestic disciplinary journals. In addition, informal communication about research is relatively infrequent and limited primarily to contact with scientists in one's own department. Scientists report that communication even within this group occurs less often than once a week. Discussions with a broader range of participants in the research system (such as scientists in other disciplines and institutions, clients, and extension staff) are limited to less than once a month and may in the future be further constrained by limited travel budgets. Despite infrequent contact with other scientists, staff, and clients, agricultural scientists do indicate that a number of these people are influential in a variety of research decisions. However, scientists in basic science departments are perceived as uniformly noninfluential for nearly all the agricultural scientists' research decisions.

Nevertheless, the nature of the specific scientific communication is closely related to the ways in which science is conducted. For example, scientists who use journals as a resource for research and an outlet for results, and communicate frequently with nonagricultural scientists, are more likely to rely on scientific and disciplinary criteria for problem choice. In contrast, those scientists who read and publish in bulletins, and communicate frequently with clients and extension staff, are more likely to be concerned with clients' needs and demands, feedback from extension, and marketable final products. Clearly, agricultural scientists are faced with forms of scientific communication and sources of influence that move them in different and potentially contradictory directions.

The formal and informal sources of scientific communication have important implications for the process and products of science. Effective agricultural research policy must address the importance of the scientific communication system, its integral relationship with the goals and products of agriculture and agricultural research, and the potential conflicts in the present system. But scientific communication is embedded in a disciplinary, an organizational, and an extraorganizational context. As psychologist M. Mahoney has observed, "in the final analysis, science involves a collection of human conjectures which are communicated in an elaborate social system and which are ultimately accepted or rejected, not on the bases of formal logical criteria, but rather in accord with the psychological and sociological principles which circumscribe our behavior as humans" (1979:366). It is to the various contexts within the larger social system that we now turn.

5
Disciplinary Influences
on Research

By the end of the first decade of this century, agricultural scientists had begun to organize themselves along both disciplinary and institutional lines. In this chapter we explore the complex differences that have developed among disciplines. Scientists in each of the disciplines have different perspectives and work within a somewhat different world, composed of a group of related research areas. These research areas are not spatial units, nor are they necessarily particular objects in the everyday world. Instead, as Richard Whitley noted, "a research area can be said to exist when scientists concur on the nature of the uncertainty common to a set of problem situations. Some agreement on what is problematic and what can be assumed is essential" (1974:80). Consequently, scientists in different research areas will see the same thing differently. For example, a corn pathologist, a corn breeder, a corn entomologist, and an animal nutritionist will each see a different scientific object when confronted with an ear of corn.

At the disciplinary level, a variety of assumptions are tacitly shared. Traditionally, each of the disciplines tends to see most—if not all—problems as resolvable through the body of knowledge and approach that that discipline offers. Other solutions are often viewed with skepticism if not outright disbelief. The situation is analogous to an attempt to see a picture on a wall in a darkened room. One may illuminate the room with a white light, a blue one, a yellow one, etc. One may even mix colors or change them in sequence. However, one does not have the option of using no color at all in an attempt to see the picture as it *really* is.

Similarly, scientists may "illuminate" the world through their respective disciplines. They may "mix colors" by engaging in interdisciplinary research, and they may "change colors in sequence" by asking colleagues in other fields how they confront a particular problem.

What they may not do is see their object of inquiry as it *really* is. The "facts" differ according to the scientist's theoretical perspective. As Einstein observed, "It is theory which decides what we can observe" (Mahoney, 1979:351). Indeed, the objects scientists see are in large part created by their disciplines (Knorr and Knorr, 1978; Latour and Woolgar, 1979). As Ravetz noted, "The objects of scientific inquiry are of a very special sort: classes of intellectually constructed things and events" (1971:109). What concerns us in this chapter, then, is how scientists in various disciplines come to see the world differently. In particular, we focus on disciplinary differences in scientists' educational background, allocation of research time, criteria for problem choice, and factors for promotion and tenure.

Educational Background

Disciplinary differences are particularly marked in the educational background of agricultural scientists. While practitioners in some fields have a broad background, those in others have had little exposure to other disciplines. In fact, the modal or most common career path in eight out of sixteen disciplines is to take all three degrees in the same field (Table 5.1). On the other hand, crop scientists, environmental scientists, food scientists, and nutritionists most frequently have taken all of their degrees in disciplines other than the one in which they practice. This diversity in educational background for these disciplines is the result of very different circumstnces. For example, most crop scientists have bachelor's degrees in general agronomy, which is often the nearest equivalent to crop science. Environmental and food science are relatively new fields; therefore, it is not surprising that few current practitioners have degrees in them. Nutritionists are often recruited from the basic sciences as well as animal science. Soil scientists most typically enter their discipline at the master's level; plant pathologists and geneticists enter at the doctoral level. This appears to be due in part to the relatively limited offerings in these fields at the lower levels.

The separateness of most agricultural disciplines is illustrated by the relatively small proportion of their practitioners' degrees that were obtained in other disciplines (Table 5.2). For example, 90% of all degrees obtained by agricultural engineers were in engineering. Among agronomists, 75% of degrees were obtained in agronomy, while 11% of degrees were in the basic sciences. A similar pattern holds for animal science and entomology. Virtually no crossover can be found between agronomy and animal science. And both agricultural economics and social science tend to have little interchange with all other fields.

TABLE 5.1
Educational Background of Scientists by Discipline

Discipline	Percent of Scientists With All Degrees in Same Discipline	Percent of Scientists With One Degree in Another Discipline	Modal Educational Path[a] Bachelors In	Bachelors Out	Masters In	Masters Out	Doctorate In	Doctorate Out	%[b]
Agricultural Economics	53	39	X		X		X		43
Agricultural Engineering	72	14	X		X		X		69
Agronomy									
All	53	26	X		X		X		45
Crop Science	8	15		X		X		X	41
Soil Science	29	37		X	X		X		29
Animal Science	56	21	X		X		X		51
Basic Sciences									
Biochemistry	6	22	X		X		X		39
Biology	32	14		X		X		X	31
Chemistry	57	4	X		X		X		36
Genetics	3	14		X	X		X		43
Entomology	41	35	X		X		X		33
Environmental Sciences	7	23		X		X		X	46
Food Science	8	26		X		X		X	45
Forestry	62	13	X		X		X		45
Horticulture	50	20	X		X		X		41
Nutrition	17	28		X		X		X	27
Plant Pathology	11	37		X		X	X		41
Social Sciences	37	35	X		X		X		33

[a] X's denote whether the modal educational path taken is within a given discipline or outside it.

[b] Indicates percentage of scientists following modal path.

TABLE 5.2
University Degrees Obtained by Agricultural Scientists, by Current Field of Science and Discipline[a]

Current Field of Science	Discipline in Which Degree Obtained (%)						
	Agricultural Economics	Agricultural Engineering[c]	Agronomy	Animal Science	Basic Sciences	Entomology	Environmental Sciences
Agricultural Economics	74	1	1	3	d	0	d
Agricultural Engineering	2	90	2	0	2	0	1
Agronomy	1	d	75	1	11	d	1
Animal Science	0	d	0	74	7	1	2
Basic Sciences	d	1	8	8	64	1	d
Entomology	0	0	2	1	12	70	d
Environmental Sciences	1	5	5	6	15	4	31
Food Science	3	8	2	18	23	0	1
Forestry	2	5	6	4	2	3	0
Horticulture	1	0	10	d	10	0	0
Nutrition	1	1	3	12	18	0	0
Plant Pathology	1	0	9	d	20	1	0
Social Sciences[b]	8	0	0	0	1	0	3

[a] Includes all graduate and undergraduate degrees; postdoctoral study is excluded.

[b] Includes Agricultural Education.

[c] Includes general engineering degrees.

[d] One-half of one percent or less.

TABLE 5.2 (cont.)

Current Field of Science	Discipline in Which Degree Obtained (%)						
	Food Science	Forestry	Horticulture	Nutrition	Plant Pathology	Agricultural Education	Social Sciences
Agricultural Economics	d	d	d	4	0	4	2
Agricultural Engineering	0	2	0	0	0	0	0
Agronomy	d	1	2	0	1	3	1
Animal Science	1	0	0	3	0	2	1
Basic Sciences	1	2	3	d	2	0	1
Entomology	0	2	0	0	0	d	0
Environmental Sciences	0	8	3	0	1	0	3
Food Science	31	0	3	5	0	0	1
Forestry	0	77	0	0	1	0	0
Horticulture	d	d	70	d	1	3	0
Nutrition	6	0	0	44	0	1	9
Plant Pathology	0	3	7	0	52	2	1
Social Sciences[b]	0	1	0	0	0	9	68

[a] Includes all graduate and undergraduate degrees; postdoctoral study is excluded.

[b] Includes Agricultural Education.

[c] Includes general engineering degrees.

[d] One-half of one percent or less.

TABLE 5.2 (cont.)

Current Field of Science	Discipline in Which Degree Obtained (%)					Total Degrees Obtained
	General Agriculture	General Science	Humanities	Zoology	Miscellaneous	
Agricultural Economics	4	1	2	0	1	307
Agricultural Engineering	1	1	0	0	1	205
Agronomy	4	1	d	d	0	745
Animal Science	7	0	0	4	0	370
Basic Sciences	1	2	d	4	d	492
Entomology	1	1	1	8	0	246
Environmental Sciences	3	1	1	15	2	156
Food Science	3	2	0	1	1	132
Forestry	0	d	0	d	d	203
Horticulture	2	0	1	0	0	240
Nutrition	5	1	0	0	0	114
Plant Pathology	3	0	d	0	0	244
Social Sciences[b]	1	0	7	0	2	130

[a]Includes all graduate and undergraduate degrees; postdoctoral study is excluded.

[b]Includes Agricultural Education.

[c]Includes general engineering degrees.

[d]One-half of one percent or less.

The basic sciences, on the other hand, tend to draw small numbers of persons from many other fields; at the same time, they provide training for essentially all the other biological fields. The environmental sciences tend to draw from the most diverse background, including the basic sciences and zoology. The background of food scientists reflects that discipline's origin as a part of dairy science, as well as the biochemical character of food research. A similar pattern holds for nutritionists.

In short, while some substantial overlaps exist, the majority of agricultural scientists obtain little exposure to fields not closely allied to their own. As one frustrated industry spokesman complained, "There has been no real attempt on the part of graduate schools to teach students how to work in a multidisciplinary team. The language is different, the method of approach to problems is considerably different, and, for all practical purposes, it is very difficult for members of a team to be able to communicate among themselves" (Bauman, 1979:30). In fact, the disciplinary societies that accredit agricultural curricula often make so many demands for disciplinary courses that even basic education in the humanities and social sciences is neglected (Kellogg and Knapp, 1966). Promising exceptions are the recently developed programs in integrated pest management (see Chapter 11).

Time Budgeting

The separate worlds of agricultural scientists become particularly evident when one examines the sharp differences among the disciplines in the allocation of research time. For example, scientists in the various plant sciences spend a quarter or more of their time out in the field. In contrast, laboratory work accounts for a quarter or more of the time of those in the basic sciences, food science, and nutrition.

Horticulturists and plant pathologists make most use of greenhouses; social and economic scientists spend a great deal of their research time in their offices (Table 5.3).

These statistical data gloss over fundamental differences in the way research is perceived and carried out in each of the agricultural disciplines. For those in the basic sciences, the definition of research corresponds to the public image of the white-coated lab scientist. On the other hand, an agricultural economist may spend three-quarters of his or her time in the office, using a desk calculator or remote computer terminal to analyze information gathered elsewhere and printed in tabular form. In the former case, the office is a place where one may go to write about the results of one's research; in the latter, the office may itself be a research site.

TABLE 5.3
Percent of Research Time in Each Location by Discipline[a]

Discipline	Office	Library	Field	Lab	Computing Facility	Greenhouse	Elsewhere
			Location				
Agricultural Economics	76	6	10	0	4	0	4
Agricultural Engineering	50	7	18	14	6	1	2
Agronomy	43	6	27	14	2	5	2
Animal Science	47	7	16	22	3	0	4
Basic Sciences	41	10	8	36	1	2	2
Entomology	39	6	23	24	2	5	1
Environmental Sciences	42	8	28	16	4	0	2
Food Science	48	10	6	33	1	0	1
Forestry	53	8	18	15	4	2	1
Horticulture	36	7	30	13	3	8	2
Nutrition	38	13	12	27	6	0	3
Plant Pathology	33	6	24	22	2	10	2
Social Sciences	67	11	14	1	3	0	3

[a]Numbers may not add to 100 due to rounding.

Differing time budgets also reflect the differential speed with which research can be accomplished. Biological field research is restricted not only by the character of the objects of study but also by the exigencies of weather, the distance necessary to reach research sites, and the wide variation found in field settings. At the other end of the spectrum, most of these constraints are eliminated in the case of soils research in a laboratory or the economic analysis of statistical data.

Time budgets may also imply certain characteristics of the researchers themselves. For example, those who "love the great outdoors" may tend to enter the agronomic/crop sciences, whereas those who find heavy labor, the hot sun, and dirt and grime unattractive are more likely to go into the basic sciences. Other fields, such as entomology and plant pathology, offer scientists the opportunity to do either field or laboratory work or both.

Scientists' time budgets appear to change over the course of their careers. In Harmon's (1965) study of doctorate cohorts in all the sciences from 1935 to 1960, he found that the distribution of time for research, teaching, and administration changed radically during one's working life. The most marked trend was the rapid growth of adminstrative responsibility, chiefly at the expense of research time. He reported that 25 years after the doctorate, administration occupied the greatest single segment of scientists' time. This trend is also present in our sample. The oldest cohort report 21.3% of their time devoted to administration and the lowest percentge of time devoted to research (52.5%) of any cohort. The cohort receiving their doctorates in the last four to eight years spend 5.7% of their time on administration and 63.3% on research. Despite the fact that all these agricultural scientists are active researchers with projects listed in the Current Research Information System, the trend toward administration for older scientists is clear. If the median age of agricultural scientists continues to increase during the next few decades, it may well reduce the quantity of time devoted to research.

Criteria for Problem Choice

Scientists from different disciplines tend to define, to choose, and to execute research in quite different ways. In part because of training and peer interaction, in part because of reward systems, and in part because of the larger social context, scientists differ markedly across disciplines in the criteria they use for problem choice. For example, confronted by the problem of low agricultural productivity in a given locale, an engineer would be likely to focus on tillage, drainage, and planting and harvesting practices; an economist might explore the availability of farm markets and credit; an agronomist would be likely to check the yield potential of the cultivars being grown and their rates of fertilization; and a sociologist might inquire into the role of the farm family as a source of labor, capital, and risk-taking behavior. Of course, any or all of these issues might be relevant to a given situation. Nevertheless, disciplinary differences would (and do) have a major impact on the way problems are chosen and defined. This is apparent from the marked disciplinary differences in the relative weight given to fifteen out of twenty-one criteria for problem choice (Table 5.4).

The single most important criterion for research problem choice, "enjoy doing this kind of research," does not differ significantly across fields. However, the second most important criterion, "importance to society," does differ, with nutritionists rating this item most highly, and those in the basic sciences giving it the lowest score. Not surprisingly, those in agricultural economics and the social sciences, whose research requires less capital expenditure, are far less concerned about the availability of research facilities than those in the biological sciences and engineering.

Scientific curiosity is ranked highest by nutritionists, who are followed by those in the basic and social sciences, fields traditionally somewhat separated from "production agriculture." Agricultural economists and agricultural engineers have the lowest scores on this item. The potential creation of new methods, useful materials, and devices is ranked high by foresters, food scientists, and agricultural engineers, reflecting the product orientation common to these fields. Those disciplines lacking in concrete products—agricultural economics and the social sciences—score lowest on this item.

Nutritionists and entomologists are most concerned about publication in professional journals ($\bar{x} = 5.3$), and agricultural economists are least concerned ($\bar{x} = 4.0$). Indeed, unlike those in all other fields, agricultural economists rate publication in experiment station or research service bulletins as far more important than publication in professional journals. Although the disciplinary means differ, only horticulturists rank publication in farm and industry journals above the midpoint on the scale.

Scientists vary in the degree to which they consider client needs, client demands, and extension feedback as important criteria for problem choice. Horticulturists appear to be the most client oriented; those in the basic sciences pay little attention to clients. However, for each of the disciplines, scientists' own assessments of client needs are

TABLE 5.4
Selected Criteria for Research Problem Choice by Field of Science[a]

Discipline	Importance to Society	Availability of Facilities	Scientific Curiosity	Creation of New Methods, Devices	Publication in Professional Journals	Client Needs As Assessed by You	Likelihood of Clear Results	Priorities of the Organization
				Criterion				
Agricultural Economics	5.6 z	3.5 x	4.4 v	4.4 v	4.0 x	5.3 yz	5.4 z	4.5 yz
Agricultural Engineering	5.8 x	5.3 z	4.5 vw	5.6 yz	4.5 xv	5.4 yz	4.5 xy	4.7 z
Agronomy	5.7 x	5.5 z	5.2 xyz	5.2 wxyz	4.9 yz	5.2 xyz	4.8 xyz	4.4 yz
Animal Science	5.7 x	5.8 z	5.1 wxy	5.2 xyz	5.1 yz	5.4 z	4.6 xv	4.4 yz
Basic Sciences	5.0 y	5.3 z	5.5 yz	4.7 wwx	5.0 yz	3.2 v	4.8 xyz	3.8 xy
Entomology	5.5 yz	5.5 z	5.4 xyz	5.0 vwxyz	5.3 z	4.8 wxy	5.0 xyz	4.6 z
Environmental Sciences	5.5 yz	5.2 z	5.4 xyz	4.9 vwxy	5.1 yz	4.6 wx	5.0 yz	4.2 xyz
Food Science	5.6 yz	5.8 z	5.0 wxy	5.6 yz	5.2 yz	4.9 wxy	4.5 xy	4.4 yz
Forestry	5.6 yz	5.2 z	5.0 wxy	5.7 z	4.6 xyz	5.2 wxyz	4.8 xyz	4.2 xyz
Horticulture	5.7 x	5.5 z	4.7 wwx	5.2 xyz	5.2 yz	5.8 z	4.5 xv	4.6 z
Nutrition	6.0 z	5.6 z	5.9 z	4.6 wwx	5.3 z	4.5 w	5.0 xyz	3.8 xy
Plant Pathology	5.6 yz	5.4 z	5.1 wxy	5.1 wxyz	5.0 yz	4.8 wxy	4.4 x	4.4 xyz
Social Sciences	5.7 x	4.2 y	5.5 yz	4.4 ww	4.6 xyz	4.8 wxy	4.9 xyz	3.7 x

[a]Mean scores based on seven point scale (1 = not important, 7 = very important). For all columns p < .005.

v-z Means within a column with different letters are significantly different, p < .01.

TABLE 5.4 (cont.)

Discipline	Contribution to Theory	Demands of Clientele	Currently a "Hot" Topic	Criterion — Marketability of Product	Publication in Bulletins	Feedback From Extension	Publication in Farm Journals
Agricultural Economics	3.1 v	4.7 yz	4.7 z	4.2 z	5.1 z	3.7 yz	3.3 y
Agricultural Engineering	3.7 ww	4.5 xyz	4.5 z	4.2 yz	3.7 vwx	3.8 yz	3.2 y
Agronomy	4.2 wxy	4.4 xyz	3.6 wxy	4.0 xyz	4.1 xv	4.0 z	3.1 xv
Animal Science	4.3 wxy	4.3 wxyz	3.6 wx	4.0 xyz	3.5 uvwx	3.9 yz	3.4 yz
Basic Sciences	5.1 z	2.9 v	3.6 wx	3.0 w	2.2 t	2.2 v	2.1 w
Entomology	4.5 xyz	4.1 wxy	4.0 wxyz	3.5 wxy	3.3 uvw	3.6 xyz	2.9 wy
Environmental Sciences	4.3 wxy	3.8 wx	3.7 wxy	3.3 wx	3.3 uvw	3.1 wxwy	2.5 wx
Food Science	4.4 xyz	3.9 wx	4.4 yz	3.9 xyz	3.0 tuv	2.8 vw	2.7 wxy
Forestry	4.1 wx	4.4 wxyz	3.9 wxyz	3.9 xyz	4.0 wxy	3.1 wxy	3.2 xy
Horticulture	3.6 vw	5.0 z	3.5 w	4.6 z	4.6 yz	4.2 z	4.1 z
Nutrition	3.6 vw	3.6 vw	3.6 wx	3.0 w	2.9 tu	2.9 vwx	2.6 wx
Plant Pathology	4.3 wxy	4.1 wxy	3.4 w	3.6 wxy	3.4 uvwx	3.7 yz	2.9 xy
Social Sciences	4.7 xyz	3.8 wx	4.3 xyz	3.8 wxy	4.0 wxy	3.6 xyz	2.8 wxy

*Mean scores based on seven point scale (1 = not important, 7 = very important). For all columns p < .005.

v-z Means within a column with different letters are significantly different, p < .01.

substantially more important than either direct client demands or extension feedback.

Agricultural engineers, followed by entomologists and horticulturists, rank the priorities of the research organization most highly. However, even for these fields, this item is relatively low on the list of criteria. Basic scientists and nutritionists are most concerned with contributing to scientific theory, while agricultural economists and horticulturists report the lowest scores on this item.

The importance of "hot" topics varies substantially across disciplines: agricultural economics, followed by agricultural engineering and food science, rank "hot" topics substantially above the mean value for all scientists. Marketability is of substantial importance to horticulturists but of little interest to those in the basic sciences.

What can we conclude, then, about the way in which disciplinary differences affect the criteria by which research problems are chosen? First, members of each discipline employ somewhat different criteria in their choice of research problems. This is to be expected, since the criteria for problem choice are integral to the process of professionalization.

Second, scientists do not see direct pressure from their peers as important for problem choice. This is evidenced by the relatively low scores given to "evaluation of research by scientists in your field," "credibility of others doing similar research," and "colleagues' approval" by scientists *in all disciplines.* In contrast, scientists are responsive to the reward structures and ideals of their disciplines, as is evidenced by the generally high scores given to "scientific curiosity" and "publication . . . in professional journals." Third, there are disciplinary differences in the scientific orientation to research problem choice. For disciplines such as nutrition and the basic sciences, scientific curiosity and potential contributions to scientific theory are important criteria for selecting research agendas.

Finally, the various disciplines differ markedly in their reliance on clients and extension staff, as well as their product orientation, in their choice of research problems. Horticulturists are most responsive, while basic scientists are least responsive to clients and extension. Food scientists, agricultural engineers, and foresters are more concerned than other agricultural scientists with the potential for creating new products.

Peer Review

The disciplines also affect research through their control of the scientific publication system. They do this, in large part, through peer review. Peer review is the favored method for selection of papers for journal publication. In contrast, bulletins are generally reviewed within the research organization in a somewhat more cursory and, undoubtedly, less uniform fashion. Review of a paper for a journal may take anywhere from two weeks to six months, and is frequently followed by requests for revisions and publication delays.

The review process and the probability of acceptance vary markedly from field to field (Lacy and Busch, 1982b). Data for each discipline are reproduced in Table 5.5. Agricultural economics and rural sociology are notable for their very low acceptance rates. Data collected by Lindsey (1978) and Beyer (1978) showed equally low acceptance rates in other social sciences and high rates in the natural sciences. It is usually argued that this difference is due, in large part, to the higher degree of agreement in the natural sciences over the central tenets of the discipline. As a result, there is less disagreement among reviewers as to what constitutes a valid contribution to the literature. Our data suggest another reason for the disparity: The low level of reliance on page charges in agricultural economics and rural sociology have the effect of limiting the number of pages available for scientific publication. Probably, both contribute to the low acceptance rates.

Regardless of their causes, low acceptance rates have a number of effects on the sciences. First, they extend the length of time between writing and publication, as papers are revised and resubmitted, or withdrawn and submitted to another journal. Second, they waste scarce secretarial resources, as papers must frequently be retyped to meet varying style requirements. Finally, they may seriously affect the promotion chances of scientists, a point to which we return in the following chapter.

High acceptance rates, however, are no guarantee that journals meet scientists' publication needs. As the volume of scientific information has grown, journals have grown ever more specialized. One scientist argued, in justification of this process:

> Plant pathologists have lacked a good publication outlet for applied and developmental studies of good scientific calibre. Perhaps the new journal *Plant Disease* will fill this need. Sophisticated journals are dominated by white coat-bearing, egotistical, basic, imperialistic researchers. These frustrate the working plant pathologist who must advance salary-wise according to the editorial objectives of these few bureaucrats.

The tendency to specialization is reflected in the development of such forums as the *Pesticides Monitoring Journal,* a journal solely devoted to the report of problems caused by pesticides. Yet, does it make social

TABLE 5.5
U.S. Scientific Journal Publishing Policies and Procedures by Discipline[a]

Discipline	Number of Journals Surveyed	Journal Information Average Number Manuscripts Submitted Annually	Average Review Time (weeks)	Percent Accepted	Percent With Page Charges	Mean Number Reviewers
Agricultural Economics	4	186	6	27	25	2.0
Agronomy	10	196	7	78	70	2.3
Animal Science	19	223	6	68	37	2.0
Basic Sciences	12	264	5	69	58	2.0
Entomology	15	126	8	81	86	2.1
Environmental Sciences	8	202	7	58	38	2.5
Food Science	5	238	4	79	40	2.0
Forestry	6	124	10	74	67	2.8
Horticulture	3	177	5	62	33	2.3
Nutrition	1	400	4	70	100	2.0
Plant Pathology	2	258	4	79	100	2.0
Rural Sociology	2	119	8	21	0	3.0

[a]No Agricultural Engineering journal editors responded to the questionnaire.

or scientific sense to report the pest-controlling effects of pesticides in one place and their "side effects" in another? Another scientist in our study deplored the proliferation applauded by the first:

> I believe fundamental insights into the nature of specific problem areas may be realized by employing analogues which bridge disciplines. Thus, Maxwell found that the fundaments [*sic*] of hydrodyamic [*sic*] theory were able to account for electromagnetic phenomena. This type of information exchange is discouraged by today's specialized journals. . . . We need to begin (once again) to support generalists who provide answers unattainable from a specialist's point of view. The wonder of DNA is that it can interpret information from a vast variety of scale levels. Today's agencies would fail to recognize DNA's capabilities because of its lack of specificity.

The information explosion in the agricultural sciences has been documented by Tidbury (1974). He suggested that there are several ways of dealing with the problem: If one is working in an obscure corner, one can ignore it and depend upon direct communication. Alternatively, one can restrict reading to a few core journals. Another strategy involves examining documentation from title-listing services. Finally, one can supplement the core journals by reading the contents lists and using abstract journals (Tidbury, 1974).

Journals as a Medium for Interdisciplinary Communication

Acceptance rates and journal review procedures reveal much about the publication activities of scientists, but they tell little about how formal information is exchanged. In order to explore how scientists cope with huge volume of formal information in agricultural research, we asked survey respondents to list up to ten journals to which they currently subscribed and up to ten journals in which they had published in the last five years. Each journal was then assigned to a discipline. In ambiguous cases, we consulted directories or examined the journal itself. We identified thirty-three separate disciplines ranging from crop science to medicine. To insure that the measure would remain conservative—i.e., that it would not underestimate interdisciplinary ties— we classified the scientists into eighteen disciplines instead of the thirteen disciplines used in the rest of the study.

Some fields have large numbers of journals with relatively few pages, and others have a few journals with many pages. Moreover, acceptance rates and subscription costs differ widely across disciplines.

Finally, there are different numbers of scientists in each discipline. In order to standardize the results across disciplines the following procedure was introduced. The number of scientists in field *i* who subscribed to or published in *at least one* journal in field *j* was determined. Then, within each field, the percentage of scientists subscribing or publishing by field was computed.

An examination of major interdisciplinary links based upon journal subscriptions reveals a number of important patterns (Table 5.6). First, the general science journals (*Science, Scientific American, BioScience*) serve as the major linkage between disciplines. In fact, those in the basic sciences are more likely to subscribe to these journals than to journals within their own disciplines. For example, biochemists are nearly twice as likely to subscribe to a general science journal as to one in biochemistry. There is one notable exception: Only 6% of agricultural economists and 7% of social scientists report subscribing to general science journals.

Second, agricultural economists and social scientists have subscription patterns divorced from those of the other agricultural sciences. Agricultural economists draw upon the social science literature to a significant degree, but only 2% of social scientists subscribe to agricultural economic journals.

A third observation concerns the direction of flow of scientific information. The pattern is clearly *not* a neat one in which the basic sciences develop knowledge that is in turn applied by the production sciences (e.g. agronomy, animal science, horticulture) and finally developed by the engineers. In fact, the basic sciences appear to rely heavily upon information derived from applied fields; for instance, a significant percentage of chemists subscribe to engineering journals.

The percentages of scientists subscribing to a journal within their own discipline varies widely across disciplines. In some fields, such as plant pathology and agricultural economics, nearly all practitioners subscribe to at least one within-field journal. In other disciplines, particularly the basic sciences, general science journals appear to be of equal or greater importance.

Finally, the relative insularity of the disciplines is reflected in the mean number of fields in which scientists have at least one subscription. Geneticists and nutritionists read most widely; they average over three disciplines each, in contrast to agricultural economists and social scientists, who subscribe, on average, to journals in less than two fields. The mean scores for agronomists may be somewhat inflated by subscriptions to journals in subfields of the discipline.

The publication patterns of agricultural scientists are quite similar to their subscription patterns, although, not surprisingly, there is less

TABLE 5.6
Percentage of Scientists <u>Subscribing</u> to at Least One Journal in a Given Field, by Field

Field of Scientist	Subscribing in Field (%)	Other Field of Subscription	Subscribing (%)	X̄ [a]
Agricultural Economics	94	Social Sciences	17	1.84
		Economics	40	
Agricultural Engineering	85	Soil Science	16	2.37
		General Science	17	
		Engineering	49	
Agronomy-Crops	86	Agronomy-General	57	2.73
		Agronomy-Soils	12	
		General Science	36	
		Botany	25	
Agronomy-General	89	Agronomy-Crops	77	2.60
		Agronomy-Soils	37	
		General Science	31	
Agronomy-Soils	86	Agronomy-Crops	19	2.70
		Agronomy-General	70	
		General Science	35	
Animal Science	89	General Science	28	2.33
		Nutrition	22	
		Medicine	15	
		Biology	16	
Biochemistry	36	Biology	26	2.56
		Chemistry	34	
		General Science	60	
Biology	31	General Science	48	2.69
		Botany	22	
		Microbiology	34	
		Animal Science	18	
		Medicine	19	
Chemistry	70	Food Science	35	2.80
		General Science	35	
		Physical Sciences	15	
		Engineering	15	
Entomology	95	Biology	17	2.02
		General Science	39	
Environmental Sciences	71	Biology	22	2.20
		General Science	35	
Food Science	91	Animal Science	30	2.80
		Chemistry	39	
		General Science	33	
Forestry	84	Biology	17	2.20
		General Science	29	
		Engineering	21	
Genetics	53	Agronomy-Crop	42	3.50
		Biology	28	
		General Science	61	
		Horticulture	56	
		Botany	28	
Horticulture	93	Agronomy-Crop	14	2.08
		Agronomy-General	21	
		General Science	20	
		Botany	20	
Nutrition	84	Animal Science	39	3.13
		Biology	37	
		General Science	47	
		Medicine	26	
Plant Pathology	97	Agronomy-Crops	23	2.45
		General Science	37	
		Horticulture	21	
Social Sciences	41	Home Economics	22	1.56
Overall Mean				2.42

[a] Mean number of fields in which scientists have at least one subscription.

TABLE 5.7
Percentage of Scientists <u>Publishing</u> at Least One Journal Article in a Given
Field, by Field, 1974-1979

Field of Scientist	Publishing in Field (%)	Other Field of Publication	Publishing (%)	$\bar{X}$ [a]
Agricultural Economics	58	Economics	16	1.35
Agricultural Engineering	73	Agronomy-Crop	17	2.17
		Agronomy-General	15	
		Engineering	36	
Agronomy-Crops	72	Agronomy-General	39	2.22
		Botany	24	
Agronomy-General	51	Agronomy-Crops	60	1.57
Agronomy-Soils	62	Agronomy-Crops	17	2.30
		Agronomy-General	61	
Animal Science	82	Veterinary Medicine	15	2.08
		Medicine	18	
Biochemistry	56	Chemistry	36	2.74
		Microbiology	20	
		Medicine	30	
		Biology	18	
Biology	25	Agronomy-Crop	16	2.46
		Animal Science	22	
		Microbiology	27	
		Medicine	31	
		General Science	15	
		Veterinary Medicine	16	
		Botany	22	
Chemistry	65	Agronomy-Crop	20	2.75
		Animal Science	20	
		Biochemistry	25	
		Food Science	30	
		Forestry	15	
		Engineering	15	
		Medicine	15	
Entomology	93	Biology	16	2.01
		General Science	18	
Environmental Science	59	General Science	27	2.06
Food Science	91	Animal Science	33	2.59
		Chemistry	41	
		Microbiology	17	
Forestry	71	Engineering	19	1.86
Genetics	50	Agronomy-Crop	39	2.92
		Agronomy-General	19	
		Animal Science	22	
		Horticulture	36	
		Botany	19	
Horticulture	81	Botany	19	1.65
Nutrition	68	Animal Science	32	2.16
		Chemistry	16	
		Food Science	16	
		Medicine	24	
Plant Pathology	92	Agronomy-Crop	26	2.41
		Horticulture	23	
		Botany	17	
Social Sciences	73	---		1.51
Overall Mean				2.26

[a] Mean number of fields in which scientists have at least one publication.

cross-disciplinary publication (Table 5.7). Scientists in the basic sciences publish in the broadest range of disciplines; those in agricultural economics and sociology publish essentially only within their respective disciplines. Plant scientists contribute significant numbers of articles to botany journals, and animal scientists contribute to medical and veterinary journals. However, few production scientists contribute to other basic science journals or to the general science journals.

In short, the subscription and publication patterns of agricultural scientists reveal not a linear flow of information, but a complex matrix of information flow. Still, most scientists read and publish only in their own or closely related fields. And certain pathways are unused: fewer than 6% of crop or animal scientists subscribe to or publish in each other's journals. Only 8% of agricultural engineers subscribe to at least one crop science journal, and an even lower percentage subscribe to an animal science journal. Finally, social scientists and agricultural economists are isolated from other agricultural scientists; they appear to be largely unaware of work in other disciplines.

Publication and Productivity

It has recently become fashionable to measure scientific productivity in terms of publication rates (Carlson, Bray, and Mountney, 1977; Salisbury, 1980). This, in itself, has no doubt contributed to the proliferation of journal articles; unlike plants and animals, scientists can consciously change their behavior in response to the measure. Nevertheless, neither the reward system nor mere time for research is sufficient to increase general levels of publication productivity.

Clearly, those with no time to devote to research will produce nothing. However, the relationship between time and productivity is curvilinear (Figure 5.1). After approximately 65% of a scientist's time is spent in research, it appears that additional increments of time do not result in an increase and may even result in a decrease in the production of journal articles and abstracts. (This appears to be the case in other fields as well; see Pelz and Andrews, 1966.) This curve suggests that there is a point of diminishing returns after which scientific productivity cannot be enhanced. Moreover, the production of bulletins, books, book chapters, and reports is unrelated to research time.

Despite the high level of importance accorded to publication as a criterion for problem choice and as a factor in tenure and promotion decisions, the distribution of research publications among scientists is quite skewed. The distribution of journal articles is less skewed than that of the other publications examined in our study (Figure 5.2).

FIGURE 5.1
Publication Frequency as a Function of Time Engaged in Research.

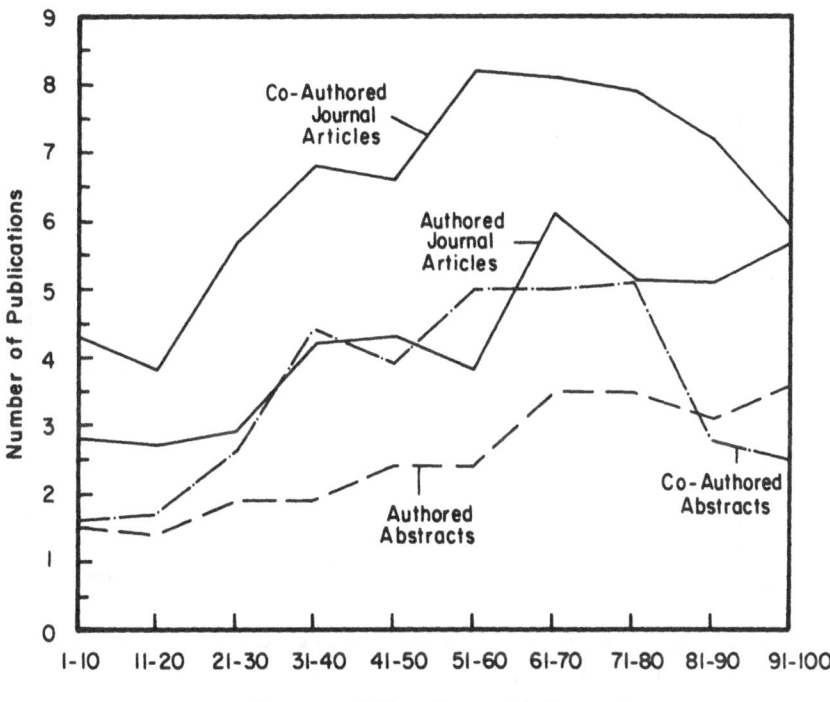

Ten percent of scientists in agricultural disciplines produced about 36% of journal articles; 50% of scientists produced 83% of articles. This figure is less skewed than that Krumland, Will, and Gorry (1979) found at the Baylor medical school. There, 10% of the faculty produced 50% of the publications. The publication of abstracts is more skewed, with 10% of scientists producing 50% of all abstracts, and 50% producing 97% of those publications.

Perhaps the most surprising finding is the even more skewed distribution for the publication of bulletins. One might expect, given the relative ease of publication in this mode, that most scientists would have published bulletins during the period in question. But the data reveal that 10% of scientists produced 62% of the bulletins, and 100% of the bulletins were produced by only 42% of the scientists. One possible explanation may be the sharp decline in the publication of bulletins at certain institutions. In other words, scientists who wish to publish may find it impossible to use this mode of publication at certain experiment stations or research facilities. Most skewed of all,

FIGURE 5.2
Distribution of Publications Among Scientists.

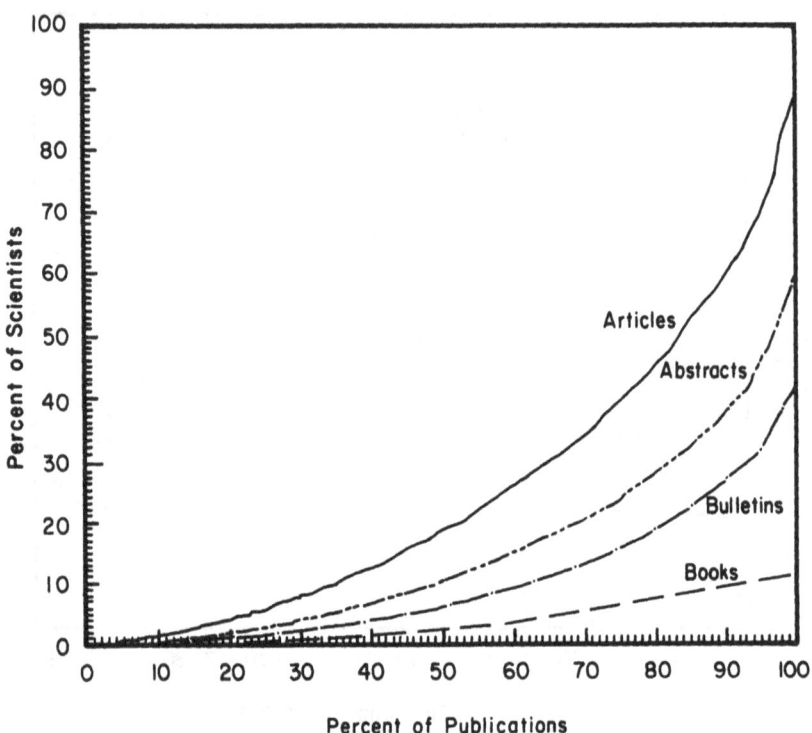

but expectedly so, is the publication of books and book chapters. Eleven percent of our respondents produced 100% of the books, and 45% produced 100% of the book chapters. This finding merely underscores the relatively minimal role played by books and book chapters in agricultural research.

In short, the majority of agricultural researchers publish relatively little. A small minority is responsible for a substantial proportion of all publications. In this respect, agricultural scientists do not differ greatly from their peers in other sciences. Although the reward structure has clearly emphasized journal publication over the last several decades, it has probably had relatively little effect on the publishing patterns of a substantial percentage of scientists.

Promotion and Tenure

Even though criteria for promotion and tenure are enforced by the organizations that employ agricultural scientists, the norms vary substantially by discipline. Scientists were asked to rank the various criteria for promotion and tenure according to their current, as well as their ideal, importance, on a scale ranging from 1 (not important) to 5 (very important) (Table 5.8). Although members of all disciplines agree that the publication of many papers is the most important criterion for promotion, food scientists see this item as most important, and social scientists as least important. Publication of high-quality papers is ranked second by those in all disciplines except animal science and nutrition. In those two disciplines, the receipt of a grant is seen as more important than the publication of high-quality papers. The receipt of a grant ranks third for members of most disciplines. No significant differences on this item among the different disciplines were found.

Among university researchers, teaching evaluations differ markedly in importance across disciplines. They appear to be least important in agricultural engineering and forestry and most important in animal science and social sciences. Equally broad differences were found on extension evaluations; not surprisingly, researchers in the basic sciences see these as least important, while those in animal science and horticulture feel that extension evaluations receive significantly higher weight in promotion and tenure decisions.

The publication of in-house reports is viewed as most important by agricultural economists, a finding consistent with their substantially higher rates of publication of bulletins. On the other end of the scale, those in the basic sciences give this item the lowest rating.

Consulting for the government is rated relatively low by members of all disciplines, with no significant differences among disciplines. Consulting with private firms, although rated relatively unimportant as a criterion for promotion and tenure by scientists in all fields, is seen as most important by food scientists and foresters.

Scientists also differ across disciplines in their perceptions of what the ideal set of criteria for promotion and tenure should be. Scientists in all disciplines agree that publication of many papers should be ranked lower than it currently is, but they differ markedly in their concern for greater quality. Food scientists perceive the greatest discrepancy between the current and ideal policies; at the other extreme, agricultural economists appear relatively satisfied.

Scientists in all disciplines wish to have the receipt of grants count somewhat less than it currently does. Differences across disciplines exist but they were relatively minor. Similarly, members of all disciplines are agreed that teaching evaluations and extension evaluations should receive more weight than they now do.

Those in the basic sciences and plant pathology would prefer that

TABLE 5.8
Criteria for Promotion and Tenure by Discipline, Current and Ideal [a]

Discipline	Criterion							
	Publish Many Papers		Publish High Quality Papers		Receipt of Grant		Teaching Evaluations[c]	
	Current[b]	Ideal[c]	Current[b]	Ideal[b]	Current[c]	Ideal[b]	Current[b]	Ideal[b]
Agricultural Economics	3.9^{yz}	3.2	3.8	4.1^{w}	3.2	2.9^{y}	3.4^{wxyz}	3.7^{wxy}
Agricultural Engineering	4.3^{z}	3.0	3.7	4.5^{xyz}	3.6	3.3^{yz}	3.1^{wx}	3.9^{w}
Agronomy	4.1^{yz}	3.2	3.8	4.4^{wxy}	3.5	3.0^{yz}	3.3^{wxyz}	3.8^{wxy}
Animal Science	4.2^{yz}	3.2	3.5	4.5^{xyz}	3.6	3.2^{yz}	3.7^{yz}	4.0^{xyz}
Basic Sciences	4.2^{yz}	3.4	3.7	4.6^{xyz}	3.5	3.3^{yz}	3.3^{wxy}	3.8^{wx}
Entomology	4.2^{yz}	3.3	3.9	4.6^{xyz}	3.7	3.4^{z}	3.4^{wxyz}	3.8^{wx}
Environmental Sciences	4.1^{yz}	3.3	3.6	4.5^{xyz}	3.6	3.5^{z}	3.5^{xyz}	4.1^{yz}
Food Science	4.4^{z}	3.5	3.7	4.8^{z}	3.9	3.5^{z}	3.4^{wxyz}	4.3^{yz}
Forestry	4.3^{z}	3.1	3.8	4.5^{xyz}	3.6	3.2^{yz}	3.1^{w}	4.1^{wxyz}
Horticulture	4.3^{z}	3.2	3.7	4.3^{wx}	3.5	2.9^{y}	3.5^{yz}	4.0^{xyz}
Nutrition	4.3^{z}	3.2	3.7	4.7^{yz}	4.0	3.4^{z}	3.6^{yz}	4.2^{yz}
Plant Pathology	4.1^{yz}	3.3	3.8	4.4^{wxy}	3.5	3.1^{yz}	3.5^{wxyz}	3.8^{wxyz}
Social Sciences	3.8^{y}	3.2	3.9	4.6^{xy}	3.4	3.2^{yz}	3.7^{z}	4.1^{z}

[a] Mean scores based on 5 point scale (1 = not important, 5 = very important).

[b] Column differences, $p < .001$.

[c] University only.

[w-z] Means within a column with different letters are significantly different, $p < .01$.

TABLE 5.8 (cont.)

| | Criterion | | | | | | | |
| Discipline | Extension Evaluations [c] | | Publish In-House Reports | | Consult for Government [c] | | Consult with Private Firm | |
	Current [b]	Ideal [b]	Current [b]	Ideal [b]	Current [c]	Ideal [b]	Current [b]	Ideal [b]
Agricultural Economics	2.9 wxyz	3.3 xyz	3.1 z	3.4 z	2.4	2.7 xv	1.9 w	2.3 wx
Agricultural Engineering	2.8 wxy	3.4 wxy	2.6 xy	2.8 xv	2.7	3.2 z	2.3 xyz	2.9 yz
Agronomy	2.9 wxyz	3.3 xyz	2.7 xyz	3.0 xyz	2.5	2.7 xv	1.9 wx	2.3 wx
Animal Science	3.2 yz	3.6 yz	2.9 yz	2.9 xv	2.5	2.8 xyz	2.2 wxyz	2.5 wxyz
Basic Sciences	2.4 w	2.5 w	2.1 w	2.0 w	2.4	2.6 x	1.9 w	2.2 w
Entomology	2.8 wxyz	3.2 wxy	2.4 wx	2.7 x	2.7	2.8 xyz	2.0 wxy	2.3 wx
Environmental Sciences	2.8 xyz	3.0 xyz	2.8 xyz	2.8 xv	2.7	3.0 yz	2.1 wxyz	2.4 wxy
Food Science	2.8 wxyz	3.2 xyz	2.4 wx	2.8 xy	2.9	3.0 xyz	2.5 z	2.7 xyz
Forestry	2.5 wx	3.2 xyz	2.8 xyz	3.1 yz	2.7	3.1 yz	2.4 yz	3.0 z
Horticulture	3.3 z	3.7 z	2.8 xyz	3.2 yz	2.5	2.7 xv	1.9 w	2.3 wx
Nutrition	2.5 wx	3.0 wx	2.4 wx	2.7 x	2.6	2.8 xv	2.1 wxyz	2.5 wxy
Plant Pathology	2.9 wxyz	3.5 xyz	2.7 xyz	2.6 x	2.4	2.5 x	2.1 wxyz	2.2 wx
Social Sciences	2.7 wxyz	3.4 xyz	2.8 xyz	3.0 xyz	2.4	2.9 xyz	1.9 wx	2.2 wx

[a] Mean scores based on 5 point scale (1 = not important, 5 = very important).

[b] Column differences, p < .001.

[c] University only.

w-z Means within a column with different letters are significantly different, p < .01.

slightly less weight be given to the publication of in-house reports. Those in animal science and the environmental sciences are satisfied with the current rankings, while those in other fields desire more weight to be given to in-house reports.

Finally, scientists in all fields wish to see further rewards given for consultation with both government and private firms. The greatest gap between the current and ideal policy on these items appears to exist for agricultural engineers.

In short, the promotion and tenure system, though a product of the experiment stations and the USDA, appears to conform to disciplinary norms. In nearly all fields, the high weight given to the publication of articles, in journals often sponsored by disciplinary associations, reinforces the role of the disciplines in certifying research results.

Short-term Research Directions

An important way in which the disciplines differ is in the topics they find to be of major current concern. To identify these topics, we asked each respondent, "What are the current 'hot' specialties within your discipline?" The major research topics that link together many of the disciplines are energy, environmental issues, and integrated pest management (Table 5.9). On the other hand, within each discipline, there is substantial divergence of opinion over the current "hot" topics. Only among agricultural engineers did more than 50% of the respondents identify the same theme, and among food scientists, nutritionists, and social scientists, no consensus emerged regarding the current "hot" topics. Therefore, in addition to the relatively sharp disciplinary boundaries, substantial fragmentation may exist even within disciplines. Further insights on this issue must await case studies of particular disciplines.

Implications

In short, agricultural scientists tend to maintain fairly sharp disciplinary divisions in their educational background, research orientation, criteria for research problem choice, and publication activities. A clear division between the plant and animal sciences as well as between the socioeconomic and natural sciences appears to exist. Moreover, longitudinal data on publications and our data on current criteria for promotion suggest that disciplinary standards have replaced organizational standards in assessing the quality of scientists' work.

This insularity among the agricultural sciences has enormous sci-

TABLE 5.9
Major "Hot" Topics Identified by Discipline[a]

Topic	Number	Percent	Topic	Number	Percent
Agricultural Economics			Environmental Sciences		
Environmental quality	31	27	Environmental quality	17	31
Energy	25	22			
International trade	15	13	Food Science		
Agricultural Engineering			(no topics identified by more than 6 persons)		
New energy forms	43	51			
Energy conservation	17	20	Forestry		
Water quality	14	17			
Environmental quality	14	17	Biomass energy	22	29
			Forest management	12	16
Agronomy			Environmental quality	10	13
Environmental quality	52	19	Horticulture		
Nitrogen fixation	42	15			
Minimum tillage	16	6	Environmental quality	12	13
Energy conservation	15	5	Tissue culture	11	12
Photosynthesis	14	5			
Genetic manipulation	14	5	Nutrition		
Modeling	13	5	(no topics identified by more than 4 persons)		
Integrated pest management	12	4			
Herbicides	11	4			
Tissue culture	10	4	Plant Pathology		
			Integrated pest management	19	21
Animal Science					
Food quality	14	11	Disease control	13	14
Protein utilization	12	9	Biological control	11	12
Reproductive efficiency	12	9	Soil pathogens	10	11
Basic Sciences			Social Sciences		
Genetic manipulation	43	23	(no topics identified by more than 5 persons)		
Environmental quality	14	7			
Tissue culture	10	5			
Entomology					
Integrated pest management	35	37			
Pheremones	14	15			
Insect behavior	14	15			
Modeling	10	11			

[a]Respondents were permitted to _write_ replies. Up to four responses were coded for each respondent.

entific and social significance. First, disciplinary problems are likely to receive more support than those that cross disciplinary lines (Ruttan, 1971). Such problems are more easily defined, easier to assess in terms of (disciplinary) significance, and more likely to contribute to (disciplinary) knowledge. But this self-reinforcement also implies that the stock of knowledge produced by each of the disciplines may become divorced from that of other disciplines.

Second, by focusing on only those aspects of the world that are deemed relevant by a particular discipline, scientists may ignore problems that lie outside their competence. As Berry suggested, disciplinary insularity may give scientists the illusion that total control over agricultural phenomena is possible: "The specialist puts himself in charge of *one* possibility. By leaving out all other possibilities, he enfranchises his little fiction of total control. Leaving out all the 'nonfunctional' or otherwise undesirable possibilities, he makes a rigid, exclusive boundary within which absolute control becomes, if not possible, at least conceivable" (1977:70). Thus, entomologists may exclude the effects of pesticides on people and animals from their research on pesticide effectiveness; or animal scientists may exclude the waste disposal issues raised by confinement feeding of animals. In short, once having divided the world for study, it may be impossible to reintegrate it.

Nagi and Corwin noted:

> Different disciplines are closed to varying degrees, depending on the characteristic balance between a discipline's internal structure and the external pressures on it. In mature disciplines researchers use their established paradigms to identify central problems and the criteria for assessing the significance of findings. . . . Scientists in disciplines still lacking formal paradigms, however, often derive their research topics from social problems, and their criteria of signific[a]nce from quasi-scientific measures (1972:17).

By these criteria, it appears that most of the agricultural sciences have matured. Yet, if such is the case, it may have come about only at the expense of the creation of a barrier between scientists and client groups. Scientific research may have become uncoupled from the everyday world of farmers and other clients, so that much disciplinary knowledge is no longer congruent with their knowledge and needs, as Minckler (1976) argued with respect to forestry research. This disconnection may partly account for the reduction in the rate of increase in U.S. agricultural productivity over the last decade and the sharp rise in soil erosion rates.

Finally, the increased disciplinary emphasis appears to be partly responsible for increased specialization on the part of farmers. Much like agricultural scientists, U.S. farmers have tended increasingly to specialize in the production of a single crop or commodity, paying scant attention to soil erosion, farm runoff, and other long-term problems. In fact, scientific agriculture has resulted not only in the adoption of scientifically validated processes and products by farmers but in the transposition of the social organization of science to farming. Thus, both science and farming have become increasingly rationalized over the last several decades. Yet, ironically, the ordering of science and farming may not be the only way, or even the most effective way, to produce food and fiber for the world. Indeed, the rationalization of agricultural production may show us the limits of rationalization itself, and of our ability to control the natural world.

6
Organizational Influences

Although scientists achieve recognition from peers in their respective disciplines, the demands of modern science require that they work within an organizational framework. From its inception, modern science has had two somewhat opposing ideals, the Cartesian and the Baconian (Haberer, 1969). As the sciences have progressed, the Baconian model has become the dominant one, particularly within the agricultural sciences. Today, capital-intensive, organized, "industrialized" science (Ravetz, 1971) is the rule. In short, little science has been replaced by big science (Price, 1963).

In this chapter, we examine some of the ways in which organizational structures influence science. Specifically, we examine three dimensions of organizational structure: the institutional context (the university as opposed to the federal government), institutional prestige, and institutional geographical location (regional differences). In each case, we look at the resources provided to scientists, their research orientation, the terms of their appointments, the criteria for promotion and tenure, and, finally, the criteria they employ in problem choice.

Contexts of Research: Universities and Government

Although the Baconian model of science as a corporate, bureaucratic structure prevails throughout the United States, there are major differences between university and government scientists. Nineteenth century U.S. scientists, including those in agriculture, frequently obtained higher degrees in Germany. Upon their return they attempted to institute two interrelated features of the German university structure: rank and academic freedom. When these ideas were introduced, however, they were substantially altered. The German professor who ruled his assistants rather autocratically was replaced by the three well-known ranks of assistant, associate, and full professor. Academic freedom, which in Germany meant the freedom of the faculty—and, particularly, the professors—to run the university, was replaced by the peculiarly U.S. notion of freedom of inquiry and speech. Today, both of these traditions have become well established in the university structure. Indeed, even within the mission-oriented structure of agricultural colleges, they play a major role. Rare were experiment station directors like John A. Myers of West Virginia, who wrote in 1889: "The organization of an experiment station, it seems to me, is similar to the organization of an army. In fact, an experiment station is simply a corps of scientific workers, together with the necessary equipment, organized for a special purpose" (1889:99). Fortunately, such individuals have been few in number. In fact, one former experiment station administrator lamented that "about the only way that new directions can be achieved is through the appointment of new staff members who . . . want to do the type of research you think should be emphasized" (Wellman, 1977:51). Similarly, one recent study of a department of agricultural economics noted that administrators tend to see their role as limited to hiring the "right" people (Cartwright, 1971).

No traditions comparable to academic freedom developed within the federal government. Like most other governmental activities, science was seen as a hierarchical, bureaucratically controlled activity under government direction. Clearly, scientists were and are afforded more freedom than other government officials. In fact, it is not easy to change research directions within USDA without staff turnover (U.S. General Accounting Office [GAO], 1977). Scientists in USDA tend to be scattered throughout the country in facilities of varying size, making autocratic control difficult. However, they do appear to operate in a far more constricted environment than their colleagues in academia (NRC, 1972). As Moore described agricultural research with USDA, "the lines of authority are clear-cut below the administrator. They go from administrator to deputy, to division directors, to branch chiefs, and on down to the smallest unit" (1967:87). On the other hand, as in the experiment stations, once a project is approved, the researcher has substantial latitude within that line of inquiry.

The Forest Service is perhaps most hierarchical in its structure. "The Forest Service is not single-minded; yet there are few expressions of individualism. Innovative thinking . . . is not encouraged. Despite the ideology of decentralization, the organization and its personnel move down the same path" (Frome, 1971:42). Indeed, over the years the Forest Service has had great difficulty coping with research findings that conflicted with official policy (Schiff, 1962).

This tendency toward bureaucratic control of scientists in USDA was reflected by our respondents. Consider the following comments:

1. Your concern about the choice and direction of research is justified. Within the [USDA laboratory] . . . chemists, physicists, and agricultural technologists are joined by a few engineers (I am one) to work on widely separated research topics. In non-chemical areas, work and topics are spread thin so that individuals work with no colleagues to compare ideas with. Supervision frequently changes discipline at each level and results in meetings and discussions in which the principal investigator is the only one who knows his work and why it should be done. When supervisors restate the work it frequently shows that they do not understand work in the discipline which they are supervising because it is not their discipline. Some research topics selected at high levels must be completely redirected by a scientist to address real problems. Some redirections become worse than the stated research.

2. It is critical that support of basic research in agriculture have continuity and freedom from bureaucracy that prevents high quality and innovative research in federally controlled programs. The hop on and off band-wagon approach taken by the Congress and administration dissipates energy and funds so that basic understanding is bypassed for collection of data that will be meaningless in five years.

3. Too much of our research is dictated from Washington and the various funding agencies of USDA. Many or most of these ideas of what is the problem to be studied come directly from Congress. . . . In my estimation these agencies are not equipped to handle problem iden-tification or research needs. They only become aware of the problem [when] it reaches the front page of the newspapers. While obviously all problem identification cannot originate with the researcher, the opposite situation, where the agency bureaucrats control and dictate what will be studied with the purse strings is unhealthy. High risk [research and problems of] questionable immediate payoff are never studied. Only close term low risk research is funded.

4. As part of a [USDA] unit doing only a little basic work and with little leeway to do work in other areas, the research directions are pretty well laid out. Actual allowable variance is quite small.

5. I have had the unique opportunity to suggest and work on my current research problem. Most others are assigned to projects chosen by higher management.

Government scientists also have greater resources at their disposal. The typical government scientist in our survey has an annual research budget of $209,000, whereas the typical university scientist has a research budget of only $68,000 ($p < 0.001$). As scientists are frequently not aware of the precise total of their research budgets, these figures should be considered as educated estimates. On the other hand, a difference of this magnitude is certainly a real one. Moreover, as government scientists seek and receive fewer grants than their colleagues in the universities, these figures reflect much higher levels of fixed annual or *formula* funding.

These figures also illustrate a problem typical of experiment stations. Low budgets *per researcher* put scientists in a rather awkward position: "This is a sort of public research institute wherein a worker receives a salary and nothing else to do research. He, then, is expected to solve a lot of agricultural problems for the state's people by boot-strapping the project's costs through his own initiative in gathering money." This encourages researchers to become research entrepreneurs; and defeats the purpose of formula funding, which is to provide adequate continuous research support.

Moreover, experiment station research is often fragmented into many small, understaffed projects. As one agricultural college dean lamented several years ago, "The greatest limitation to under-financed marketing research is the 'shotgun approach' which we have used where we have too many projects, some with low priority, understaffed and underfunded" (Ewing, 1975:3). Clearly stemming in part from the desire to respond to the demands of diverse clientele, the "shotgun approach" appears not to be limited to marketing research. Consider, for example, these comments by scientists responding to our survey:

> The state needs to hire fewer people at the agricultural experiment stations and support them better, focusing their goals, and making sure that they are in communication with the people.

> The _____ State Legislature has never properly funded either teaching or research at this institution. Teaching is not funded beyond salaries; thus it is necessary for teaching to parasitize research funding. The legislature does not fund a single full-time technician in the whole College of Agriculture.

The hierarchical character of government research is also reflected in its support staff (Table 6.1). While university scientists have substantial numbers of graduate assistants, government scientists have a greater number of technicians and "other staff." Furthermore, graduate students are in the process of being trained and generally work no more than 20 hours per week; technicians are already trained and are generally full-time staff members. A similar case can be made for "other staff"; university scientists are far more likely to hire undergraduates as part-time workers in their laboratories and fields, whereas government scientists are more likely to hire full-time staff on civil

TABLE 6.1
Mean Number of Support Staff per Scientist of Government
and University Scientists

	Institution	
Staff	Universities	Government
Graduate Students	2.6[a]	0.5
Postdoctorates	0.2	0.2
Technicians	1.5[a]	2.5
Other Staff	0.8[a]	2.3

[a]Row differences, p < .001.

TABLE 6.2
Actual Terms of Appointment for University and Government Scientists[a]

	Institution	
Terms (%)	Universities	Government
Research	57.4	77.1
Teaching	27.0	0.2
Administration	8.5	20.0
Extension	6.3	2.0

[a]Totals do not add to 100 due to small percentage of
time spent in "other" activities. For all row differences,
p < .001.

service payrolls. If we assume that all graduate students work half
time and that "other staff" at universities work two-thirds time, but
those in government work full time, we arrive at an estimate of 3.53
and 5.25 full-time equivalent staff members for each university scientist
and government scientist, respectively.

In addition to having a larger support staff and budget, government
scientists spend far more time doing research. University scientists
spend 57% of their time in research, while government scientists
spend 77% of their time in research (Table 6.2). On the other hand,
the more bureaucratic nature of government science is illustrated by
the fact that, whereas university scientists spend 8½% of their time
in administrative tasks, their colleagues in government spend 20% of

TABLE 6.3
Perceived Criteria for Promotion and Tenure for University and
Government Scientists, Current and Ideal[a]

Criterion	Current		Ideal	
	University	Government	University	Government
Publish many papers	4.1	4.2	3.2	3.3
Publish high quality papers	3.7[b]	4.1	4.5	4.5
Publish in-house reports	2.7[b]	2.4	2.9[b]	2.6
Consult with government	2.5[b]	2.8	2.7[b]	3.1
Consult with private firms	2.0	2.1	2.3	2.5
Receipt of grant or contract	3.8[b]	2.4	3.3[b]	2.4
Teaching evaluations	3.4	--	3.9	--
Extension evaluations	2.8	--	3.2	--

[a]Mean scores based on five point scale (1 = not important, 5 = very important).

[b]Row pair differences, p <.01.

their time in administration. Moreover, many university faculty are simultaneously engaged in research, teaching, and extension activities; government scientists tend to focus exclusively on research. And, despite official pronouncements to the contrary (USDA, Special Work Group, 1977), government scientists report doing the same amount of basic research as their colleagues in universities.

Not surprisingly, the greater financial and human resources and research time available to government scientists lead to substantially higher publication rates. During the period from 1974 to 1979, university scientists published an average of 4.3 single-authored journal articles each. Their colleagues in government, on the other hand, published 6.1 such articles ($p < 0.005$). Importantly, although government scientists published nearly 50% more single-authored journal articles, there is little difference between government and university scientists in the publication of multiple-authored journal articles. This means that, as measured by publications, USDA research is much more expensive than experiment station research (see also NRC, 1972).

Criteria for promotion and tenure also differ for university and government scientists (Table 6.3). Both government and university scientists see the publication of *many* papers as the most important criterion currently employed in promotion decisions. There is no significant difference between government and university scientists on this criterion. The publication of high quality papers is perceived to be less important in promotion and tenure decisions for those in universities than for those in government. In contrast, Warner and Bruce (1975) found that experiment station directors saw publication in general as more important than did researchers, but also saw quality as more important than quantity (see also Heady, 1972; Magrabi, 1969).

The publication of in-house reports, though less important than the publication of papers, is of greater importance to the scientists in universities than to their colleagues in government. This is reflected in the nonsignificant difference in the rate of publication of in-house reports by government and university scientists despite the greater resources of those in government.

Consulting with either government or private firms ranks relatively low on the list of criteria, though consulting for government is, not surprisingly, of greater importance to scientists in government service. Receipt of a grant or contract is extremely important to university scientists—ranking second among the various criteria—but far less so to their colleagues in government service. Finally, teaching evaluations play a fairly significant role in university promotion decisions, and extension evaluations, a relatively limited one. As one respondent succinctly stated: "In any horticulture program there is a large amount of service type work which is very important to the farmers and agribusiness of the state. However this type of work is minimized in the rewards system and one must avoid it to survive." Neither teaching nor extension evaluations plays any role in promotion decisions for government scientists.

Importantly, both university and government scientists wish to see the publication of *many* papers play a far lesser role in promotion and tenure decisions than it currently does (Grierson, 1980). In fact, several scientists saw the strong emphasis on quantity as having adverse effects on the sciences:

"Publish or perish" is not just a myth; it is very real. Administrators demand that scientists publish prolifically. In my discipline of fruit breeding, the major effort is to develop new cultivars useful to the fruit industry. Academically, the introduction of new, useful cultivars yields no "goodie points." Thus, research to introduce new fruit cultivars is counterproductive to publication.

Nevertheless, it is likely that in the foreseeable future publication of many journal articles will remain important in promotion and tenure decisions. Both demands for accountability and the ease with which journal publications can be counted will, doubtless, contribute to this situation. (At least one poultry research planner uses citation counts and publication tallies as a decision-making tool; see Hadwiger [1982].)

University scientists also wish to see the receipt of grants or contracts play a lesser role than is currently the case. In fact, many university scientists feel that the pressure to obtain research grants is so strong that it seriously detracts from the research enterprise. One scientist noted:

> I strongly feel that so much attention has been paid to funding, to grants and contracts, that the fundamental purpose of scientific research—to increase organized knowledge and how to use it —has suffered greatly. I don't hear colleagues talk of break-throughs or strong inference anymore, only about "this wonderful new program" or the "number of Ph.D.s we've turned out" or "how we plan to expand into this or that." The specific objective of the research is seldom mentioned, and when it is, it sounds like more of the same old language: "Improve our understanding of . . . ," "Study the relationship between . . ." and "Build a model to account for. . . ." All vague, all open-ended—not a specific goal in sight. *Program funding* has become the objective of scientific research! Finally, I think too much money can ruin good research. The good researcher, like the good preacher, needs to run a little hungry. The current over-funding has sucked a lot of hack-workers into the scientific enterprise. Ambitious in the most ordinary sense, they are constantly "discovering" that which wasn't lost, a strategy that gets them profitable attention but at the same time keeps raising the noise-to-signal ratio in science.

Both university and government scientists wish to see greater emphasis on all other criteria listed. In particular, both university and government scientists wish to see the publication of high quality papers count as the most important criterion for both promotion and tenure. In short, most agricultural scientists see the increasing emphasis on the production of large numbers of journal publications as undesirable.

University and government scientists differ on a number of criteria they employ in research problem choice. In fact, on eight out of twenty-one criteria, significant differences between university and government scientists appear. These differences tend to reflect once again the hierarchical character of government research as well as a stronger product orientation on the part of government scientists. For example, we found that government scientists see "enjoyment"

TABLE 6.4
Selected Criteria for Research Problem Choice by Type of
Institution[a]

| | Institution | |
Criterion	Universities	Government
Enjoy doing this kind of research	5.9	5.4
Scientific curiosity	5.2	4.6
Creation of new methods, materials, devices	5.0	5.3
Priorities of the research organization	4.1	5.3
Marketability of the final product	3.7	4.0
Colleagues' approval	3.7	4.0
Publication in bulletins	3.7	3.3
Feedback from extension	3.5	3.2

[a]Mean scores based on seven point scale (1 = not important,
7 = very important). For all row pair differences, $p < .01$.

and "scientific curiosity" as less important criteria for problem choice than do their colleagues in universities (Table 6.4). On the other hand, they see the "priorities of the research organization" and their "colleagues' approval" as of much greater importance than do university scientists. In addition, the "creation of new methods, materials, and devices," as well as "marketability," are considered more important by scientists in government than by those in universities. This is of particular importance given the official policy of focusing on more basic, long-term research in government laboratories. Finally, both "publication in bulletins" and "feedback from extension" are less important for government scientists, a difference that reflects the lower rewards that such activities receive, as well as the formal separation of research and extension activities in government service.

In sum, government scientists receive greater resources, devote more time to research, publish more, and are evaluated more in terms of research contributions than their colleagues in universities. On the other hand, government scientists tend to work in a much more hierarchical organizational structure in which organizational priorities, the creation of concrete products, and marketability figure to a significantly greater degree in the conceptualization and execution of research problems. Indeed, it may well be that the very hierarchical nature of government research tends to encourage scientists to focus upon relatively narrow, concrete products rather than the more fundamental, long-term issues that are more difficult to conceptualize unambiguously.

Institutional Prestige

In order to distinguish levels of prestige among universities, we used the somewhat controversial prestige scores generated by Ladd and Lipset (1979). Our methodological appendix provides a summary of the substantial criticisms that have surrounded the methods utilized by Ladd and Lipset to generate these scores. A detailed account of the controversy can be found in Lang (1981). Prestige scores are based upon the percent of faculty sampled who considered a particular institution to be the best in its area. Therefore, the scores we used reflect differences among colleges of agriculture, in general, rather than differences within each particular discipline. In order to simplify presentation, we have reported only mean scores of the scientists at the most and least prestigious institutions herein.

Somewhat surprisingly, scientists at more prestigious institutions have research budgets and research time allocations almost identical to those of their colleagues at less prestigious institutions, although the most prestigious institutions tend to be somewhat larger than those that are less prestigious. However, scientists at the most prestigious institutions tend to have both more graduate students (3.0 versus 2.3; $p < 0.0001$) and more postdoctoral associates (0.6 versus 0.1; $p < 0.0001$) than those at the least prestigious institutions. Such differences are no doubt attributable to the somewhat larger graduate programs at the more prestigious institutions.

The research orientations of scientists differ markedly across prestige categories. Those in the low prestige institutions spend 28% and 59% of their time engaged in basic and applied research, respectively. In contrast, those at the most prestigious institutions employ 34% of their time in basic research and 53% of their time in applied research ($p < 0.005$). Percentages of time spent on development do not differ significantly across prestige categories. Moreover, although the high prestige institutions tend to have somewhat greater proportions of their faculty in food science, nutrition, and the social sciences, these differences in disciplinary emphasis cannot account for their basic research focus.

Despite the relatively minimal differences in the resources available, there are marked differences in the average level of publication between

TABLE 6.5
Mean Five Year Rates of Publication by Prestige of
Institution, 1974-1979

	Prestige	
Type of Publication	Least Prestigious	Most Prestigious
Journal articles		
authored	3.9$_a$	6.1
co-authored	5.8^a	9.0
Books		
authored	0.1$_a$	0.2
co-authored	0.1^a	0.3
Book chapters		
authored	0.5^a	1.0
co-authored	0.3^a	0.7
Abstracts		
authored	2.7	2.7
co-authored	3.5	4.5
Bulletins		
authored	1.1	1.6
co-authored	1.1	1.1
Reports		
authored	4.1	3.2
co-authored	2.6	2.6

^{a}Row differences, $p < .01$.

those persons at the most prestigious and those at the least prestigious institutions (Table 6.5). In fact, the average publication rates for scientists at the most prestigious institutions are far above those of scientists in government, despite the much greater resources available to the latter group. However, it should be noted that most of the differences in publication rates are not statistically significant. This suggests that the variance in productivity among scientists at the most prestigious institutions is much greater than at the least prestigious institutions. The most prestigious institutions are distinguished by having a small number of highly prolific scientists. The modal or typical publication rates of scientists appear not to differ greatly across prestige categories. However, those few high producers are enough to make a very significant impact on the prestige and visibility of particular institutions within the scientific community. This suggests that agricultural scientists follow a more general pattern described by

TABLE 6.6
Selected Criteria for Promotion and Tenure of University
Scientists by Prestige of Institution[a]

| | Prestige | |
Criterion	Least Prestigious	Most Prestigious
Publish high quality papers		
Current policy	3.6^b	3.9
Ideal policy	4.5	4.5
Publish in-house reports		
Current policy	2.8^b	2.6
Ideal policy	3.0^b	2.6
Consult for government		
Current policy	2.5_b	2.5
Ideal policy	2.8^b	2.5
Consult for private firms		
Current policy	2.1_b	1.9
Ideal policy	2.5^b	2.2

[a]Mean scores based on five point scale (1 = not important,
5 = very important).

[b]Row differences, p < .001.

Ravetz (1971), in which a class system has developed within science,
and leading scientists are concentrated at a few institutions.

The reward systems of the most prestigious institutions are significantly different in a number of ways from those of the least prestigious
institutions. Moreover, scientists' perceptions of how those reward
systems should ideally function differ in important ways across prestige
categories. Publication of high-quality papers apears to be rewarded
more at the most prestigious institutions (Table 6.6), although the
rewards for the quantity of papers published do not differ across
prestige categories. However, scientists in all prestige categories would
like to see the publication of high-quality papers play a more important
role in promotion and tenure decisions.

In contrast, scientists at the least prestigious institutions are rewarded
more for publishing in-house reports. Moreover, when asked how
important they felt various criteria for promotion and tenure ought
to be, scientists at the least prestigious institutions were likely to rate
the publication of in-house reports and consulting for government
and private firms significantly higher than their colleagues at the more

TABLE 6.7
Selected Criteria for Problem Choice of University Scientists
by Prestige of Institution[a]

| | Prestige | |
Criterion	Least Prestigious	Most Prestigious
Publication in professional journals	5.0	4.2
Client needs as assessed by you	5.0	4.5
Priorities of the organization	4.4	3.5
Demands of clientele	4.3	3.7
Currently a "hot" topic	4.0	3.5
Time required to complete research	3.9	3.2
Marketability of final product	3.9	3.3
Publication in bulletins	4.0	2.9
Feedback from extension	3.7	3.4
Publication in farm journals	3.2	2.4

[a]Mean scores based upon seven point scale (1 = not important, 7 = very important). For all row differences, $p < .01$.

prestigious universities. No differences were found in the role of research grants, teaching evaluations, or extension evaluations across prestige categories. Thus, one is led to conclude that somewhat different factors enter into promotion and tenure decisions at the low-prestige institutions than at the high prestige institutions. In addition, scientists at the least prestigious institutions have a somewhat different set of organizational ideals than their colleagues.

Prestige differences are also reflected in the differing importance ascribed to various criteria for research problem choice by scientists at the most and least prestigious institutions. In fact, significant differences appear on ten out of twenty-one criteria for problem choice (Table 6.7). Importantly, in each case, scientists at the least prestigious institutions have higher mean scores than their colleagues at the most prestigious institutions. Those at the least prestigious institutions are likely to give more weight in decisions about problem choice to criteria concerning all kinds of publication, to client needs and demands, to organizational priorities, and to relatively short-term research projects as evidenced by the relatively high scores given to time constraints, marketability, and "hot" topics.

Perhaps of equal importance is the fact that scientists at the least prestigious institutions do not differ from their colleagues at more prestigious ones on those criteria relating to science ("potential contribution to scientific theory") or on more general criteria such as "enjoyment" and "importance to society." Thus, it appears that those at the least prestigious institutions tend to take the greatest number of criteria into account in choosing research problems. Those at the high prestige institutions, although no more "science-oriented" than their colleagues at low prestige universities, are less concerned about extrascientific criteria in their choice of research topics, as well as in the ways in which they are evaluated for promotion and tenure.

Prestige seems to be inversely related to the mission orientation of the agricultural colleges. Scientists in high prestige institutions are less likely to view client needs and direct applications of their research as important considerations in selecting problems than those in low prestige universities; those in high prestige institutions are also more oriented to basic research.

Regional Differences

The large size of the United States, its relative cultural diversity, and the relative autonomy of each of the State Agricultural Experiment Stations suggest that regional differences in organizational structure might exist. Moreover, though we have been unable to find any printed documentation, informants tell us that southern institutions tend to be both more hierarchical and more short-term in their approach to research problems. Certainly, institutions in the southeast have traditionally been less well funded than comparable institutions in other parts of the country, a difference that reflects regional income disparities.

In order to examine these issues, we divided experiment station scientists into four groups corresponding to USDA experiment station regions and analyzed their responses to our survey. We found the differences to be far less than expected. For example, there are no regional differences in per-scientist research budgets. Similarly, no differences in the percentage of time spent in basic research, applied research, or development are apparent. However, we did find significant differences in the mean number of support staff per scientist across experiment station regions. In the case of both graduate students and postdoctorates, the north central region reports the highest levels of support (3.0 and 0.3, respectively) while the southern region reports the lowest (2.2 and 0.1, respectively; $p < 0.01$). No differences, however, appear for technicians and other staff. These relatively small

TABLE 6.8
Selected Criteria for Promotion and Tenure of University Scientists
by Location of Institution[a]

| | | Region | | |
Criterion	Northeast	North Central	Southern	Western
Publish many papers				
Current policy	4.3[b]	3.9	4.2	4.2
Ideal policy	3.2	3.2	3.3	3.3
Teaching evaluations				
Current policy	3.4[b]	3.6	3.1	3.6
Ideal policy	4.0	4.0	3.8	3.8
Extension evaluations				
Current policy	2.9[b]	3.0	2.8	2.6
Ideal policy	3.5[b]	3.4	3.1	3.0
Publish in-house bulletins				
Current policy	2.6[b]	2.6	2.9	2.6
Ideal policy	2.9[b]	2.7	3.1	2.7
Consult with private firms				
Current policy	2.1[b]	2.1	2.1	1.9
Ideal policy	2.2[b]	2.2	2.6	2.2

[a]Mean scores based on five point scale (1 = not important, 5 = very important).

[b]Row differences, p < .01.

differences in resources available across regions are reflected in the almost identical publication rates for all kinds of publications across experiment station regions.

Important regional differences do appear in the criteria employed for promotion and tenure (Table 6.8). The publication of many papers appears to be most important in the northeast region and least important in the north central region. However, scientists are in agreement across regions on the need to reduce significantly the importance given to this criterion in promotion and tenure decisions, and to increase the importance given to the publication of high quality papers.

Teaching evaluations are most important in the north central and western regions and least important in the southern region. However, here again, scientists agree across regions that teaching evaluations should receive more weight than they currently do. Extension evaluations appear to be most important as a factor in promotion and tenure to those in the north central region and least important to those in the West. Scientists in all regions would prefer to see more

TABLE 6.9
Selected Criteria for Problem Choice of University Scientists by
Location of Institution[a]

| | | Region | | |
| | | North | | |
Criterion	Northeast	Central	Southern	Western
Publication in professional journals	4.7[b]	4.6	5.0	5.1
Client needs as assessed by you	4.6[b]	4.8	5.2	4.8
Priorities of the organization	4.0[b]	3.8	4.4	4.0
Marketability of final product	3.8[b]	3.3	3.9	3.6
Publication in bulletins	3.5[b]	3.4	4.0	3.7
Feedback from extension	3.5[c]	3.4	3.8	3.4
Publication in farm journals	3.0[c]	2.7	3.2	3.0

[a]Mean scores based upon seven point scale (1 = not important, 7 = very important).

[b]Row differences, $p < .01$.

[c]Row differences, $p < .05$.

weight given to extension evaluations, but those in the northeast give the highest ideal score to this criterion.

The publication of in-house bulletins is currently of greatest significance to those in the southern region, and southern scientists would prefer that it receive even greater consideration than it now does. Finally, consulting with private firms is relatively unimportant for scientists in all regions. Only scientists in the South desire a substantial increase in the importance of this criterion in promotion and tenure decisions.

The southern region also emerges as distinct in its ranking of criteria for problem choice (Table 6.9). Southern scientists scored highest on six out of seven criteria for which there are regional differences. They are more concerned with clients, organizational priorities, and publications of all kinds as criteria for research choices than their colleagues in other regions, although they do not differ from their colleagues in their relative concern for disciplinary or scientific interests.

Conclusions

The organizational structure of public agricultural research in the United States is an extraordinarily complex and diverse one. Scientists in different organizational settings employ somewhat different criteria for problem choice and are rewarded in different ways. Apparently government scientists work within the most constrained situation. They have less choice in their selection of research problems, carry a narrower range of nonresearch responsibilities, and are evaluated by a somewhat narrower set of criteria. Among the universities, institutional prestige appears to determine the relative importance of extrascientific constraints on research problem choice, as well as the degree of graduate student and postdoctoral support. University prestige is also reflected in *average* publication rates, although not modal ones; a small number of scientists at the most prestigious institutions produce far more than their share of research publications. Finally, southern institutions are somewhat more hierarchical and probably more short-term in their approach to research problems than those in other regions. However, these differences are far smaller than we had expected.

One theme of particular importance emerges from this examination of organizational structure in agricultural research: the need for extramural funding for research. University scientists, in particular, agree that there is too much pressure to obtain outside grants. Our data reveal that USDA research is far better supported in terms of research staff, budgets, and time available than that within experiment stations. As there is little likelihood that research funding will improve significantly in the near future, this means that the search for extramural support, and the influences it brings, will be a major issue in the coming decade. It is to that issue that we now turn.

7
Extraorganizational Influences

Agricultural scientists engage in a wide variety of extraorganizational activities in order to increase their research support, exchange information, increase their visibility and prestige, and, not least, increase their incomes. Such activities range from seeking grants or contracts to consulting and participating in review panels and advisory groups. Scientists engage in these extraorganizational activities through a wide range of organizations, including a multitude of federal agencies, state and local governments, and various private organizations. Scientists' motivations for participating in these activities and the influences that these activities have upon their research vary widely depending upon both the social location of the scientist and the particular characteristics of the extraorganizational tie.

The organizations that provide grants, contracts, or consulting opportunities all have their own interests. Public agencies have legislated mandates to work within particular domains. These may be very broad, as in the case of the National Science Foundation, or relatively narrow, as in that of the Environmental Protection Agency.

A similar range of mandates exists among private organizations. Foundations often have broad missions such as the promotion of higher education or the advancement of science. Commodity associations and private corporations are generally more narrowly oriented toward a single commodity or group of commodities. Moreover, corporations and commodity associations, generally, are interested in research only on those aspects of a commodity that are likely to lead to a higher private rate of return. These extraorganizational ties are reciprocal; both the organization and the researcher establish such ties in order to reap certain perceived benefits.

The extraorganizational ties, however, may involve conflicts of interest and important political and ethical issues regarding the nature and direction of research. For example, agricultural economist Vernon Ruttan (1982b) has observed the growing concern about linkages and boundaries between public and private sectors with respect to agricultural research. He noted that populist critics of the public agricultural research system have suggested that "private sector objectives have been given too much weight in the selection of public sector research portfolios. Official critics have argued that public research in such fields as harvest technology and mechanization belongs in the private sector and is wasteful of public resources" (1982b:23). The position being taken is that private industry should pay for this highly applied and often profitable research from its own funds. A further concern regarding the private-public linkage has been the question of who gains and who loses as the result of certain research. Because there is no guarantee that private profits will be consonant with the public good, such relationships raise a number of ethical issues.

This concern prompted former Secretary of Agriculture Bergland to state, "We will not put federal money into research where—other factors being equal or neutral—the major effect of that research will be the replacing of an adequate and willing work force with machines" (Marshall, 1980:579). Some industry representatives have been critical of state and federal research linkages with private sector research on product development and have suggested that resources devoted to product development could be used more effectively for more basic research. In contrast, Emil Mrak, chancellor emeritus of the University of California at Davis, has argued that the real threat to good agricultural science and economics has come from consumerism and its influence on research scientists in the public sector. Recently, he concluded that the USDA has been invaded by disciples of Ralph Nader and has been converted to the department of consumer culture (Marshall, 1980). In sum, extraorganizational relationships raise numerous scientific questions as well as social, political, and ethical issues.

In the past, some organizations have sought to intervene directly in public agricultural research by lobbying for funds for a particular commodity, by providing funds for specific work, and by hiring. In extreme cases various organizations have sought to prohibit or to discourage research that might lead to results unfavorable to their special interests. Some of these interventions have been documented in Chapter 1 (see also Fiske, 1978; Friedland, 1979; Hadwiger, 1982; Lacy and Busch, 1982a). Researchers in those sciences lacking organized support groups have been particularly vulnerable to such actions. Such interventions have been relatively rare when the system as a whole is considered, and might easily be rationalized as unfortunate blemishes on an otherwise admirable record. However, their influence extends far beyond the cases themselves: "Through the punishment of people who raise critical or embarrassing issues or who produce data that undermines established and institutionalized relationships, a normative climate is established. . . . That climate makes clear which topics are controversial, difficult, not-to-be-funded, and therefore to be avoided" (Friedland, 1979:10). For example, even today there are few farm labor studies being conducted in California and Florida—two states with very substantial farm worker populations and problems. More recently, the Human Nutrition Center at USDA was dismantled, and nutrition research downplayed, in large part because of organized opposition to its findings by animal agricultural interests (Broad, 1981). As of early 1981, total staff at the three new USDA nutrition centers was six (OTA, 1981).

These potential conflicts were recognized by research administrators at least fifty years ago, when few private organizations made grants to public research institutions. At that time, the issue of external ties was considered sufficiently important to warrant general policy guidelines for all the experiment stations. Both "commercial" research grants and "outside work" were the subject of a report of the Experiment Station Committee on Organization and Policy in 1927 (Barre et al., 1928). Four conditions were recommended as guidelines in determining whether or not to accept grants from "commercial" organizations. First, the research was to be of general public importance. Second, funds were to be made available through institutions and not directly to scientists. Third, a detailed agreement was to be drawn reserving all patent rights to the public institution. Finally, results were to be made public through regular channels and to the granting agency whether or not they were favorable.

It was also recommended that "outside work" be even more circumscribed. The acceptance of fees was frowned upon, except in special cases when the terms were carefully specified. Only compensation for publication was exempted from these caveats.

In the post–World War II period, the volume of private monies received by the land-grant universities began to increase markedly. Many universities created research foundations in order to smooth the transfer of such funds. As an observer noted at the time, "The role of these foundations is the promotion and encouragement of close cooperation between universities and industries, both in the training of men for industrial research, and the actual solution of problems of mutual interest and concern" (Olpin, 1944:126). At some institutions, general policies were formulated limiting projects to those that could have been financed by public funds if such funds were available (Guterman, 1944). At other institutions, doubtless, much looser guidelines permitted any kind of research as long as overhead expenses were paid.

Over the years, the proportion of scientists receiving grants, contracts, or consulting fees has increased significantly. Between 1966 and 1979, private grants at the experiment stations increased in constant dollars by 63% (OTA, 1981). Unfortunately, it has proved extremely difficult to assess the role that private grants, contracts, and consulting play in public agricultural research. Hightower (1973) saw the very fact that agricultural colleges received grants from private corporations as evidence of a conspiracy against small farmers and the general public. Agricultural economist Harold Breimyer has argued that "the most obvious challenge to the integrity of the university and its scholars is private funding of research, either institutionally or as private consulting. These are alike in that both compromise the basic role of the land-grant university. That role is to spread knowledge, to make it a public good" (1973:995). Defenders of the current system have countered that such grants are usually small and merely serve to supplement ongoing projects.

It appears likely, however, that most scientists have not seriously examined the ethical issues. Alex McCalla, former dean of the University of California at Davis, observed:

> Many people in the agricultural research establishment have grown up with this association so that to a considerable extent those inside the system share the same values as clientele groups. Therefore, they implicitly identify with their objectives. It is by this access more than by any other that commercial agriculture has had, and continues to have pervasive influence on the scope of agricultural research (1977:21).

Willard Fifield, director of the Florida Agricultural Experiment Station, had a similar insight:

> Fundamentally, public and private research workers are the same kind of people. They grew up together and were educated in the same kinds of institutions (1954:17).

And Breimyer argues:

> The more general threat to the integrity of university scholars is not that which is bought but that which is subtly induced. Some research and extension economists attach themselves to an interest group so tightly as almost to be indentured (1973:995).

Indeed, the relatively unproblematic way in which industry ties are created is underscored by a study of horticulturists in which 77%

preferred industry support for research projects (Barham, 1977), and by our own study in which scientists see agribusiness as nearly as likely to benefit from their research as farmers or the general public (see Chapter 8).

The current institutional climate in both experiment stations and USDA favors an acritical acceptance of private funds. As we noted in Chapter 6, reward systems strongly encourage the pursuit of extramural funds. Bredahl, Bryant, and Ruttan (1980) argued that the individual scientist is likely to seek such support only when his or her needs for research funds cannot be met with institutional funds. Under increasingly restricted budgets, annual appropriations tend to be channeled more and more into salaries and operating expenses, and funds for research become increasingly scarce. "However, to do a specific research project, additional 'flexible' resources are required. The agency or group that provides the marginal addition of 'spendable cash' can influence the entire research activity including substantial magnitudes of permanent state and federal funds" (McCalla, 1977:20–21). In some cases, scientists may deliberately distort research results in order to satisfy funding sources. It is more likely, however, that scientists will avoid certain topics that might be poorly received by the funding source. As annual appropriations decline, administrators may well discourage certain kinds of research for fear that grantors will take their funds elsewhere. With this context in mind, we reviewed some of the types of formal extraorganizational linkages that currently exist among agricultural scientists. (The use of a random sample precludes obtaining information on certain potentially important types of linkages; for example, Hadwiger [1982] reports that land-grant administrators frequently serve as directors of agribusiness corporations.)

Types of Linkages

Not surprisingly, USDA occupies a central position among the major types of federal extraorganizational linkages to agricultural scientists (Table 7.1). Even scientists within the Science and Education Administration report that they occasionally acted as consultants for other parts of the Department of Agriculture. The National Science Foundation (NSF) and the National Institutes of Health (NIH) also appear to be important sources of grants for public agricultural scientists. In comparison, most other federal agencies provide very few grants or contracts to agricultural scientists. This almost undoubtedly reflects a frequently made assumption within governmental circles that support for agricultural research is essentially the province of USDA. As one

TABLE 7.1
Scientists' Extraorganizational Linkages by Type of Relationship, 1974-1979[a]

| | Type of Relationship (%) | | | | | |
Organization	Grant	Contract	Review Panel	Advisory Group	Regular Consultant	Occasional Consultant
Federal Government						
U.S. Department of						
Agriculture	34	18	13	11	5	16
National Science						
Foundation	9	2	6	2	1	6
National Institutes						
of Health	8	2	2	1	1	3
Other federal unit	9	7	4	3	2	6
Other						
Foundation	12	3	2	2	2	5
Nonprofit						
organization	8	4	2	3	2	6
Publisher	2	1	1	2	2	9
Commodity						
association	10	4	3	4	4	5
Private						
corporation	15	10	1	2	9	19
State government	15	10	4	9	6	11
Local government	3	3	1	3	2	7
International						
organization	3	1	1	2	1	8

[a]Small numbers of scientists reported relationships with the Departments of Defense, Interior, and State as well as the Food and Drug Administration and the National Aeronautics and Space Administration.

researcher observed, "NSF looks down their nose at agricultural projects." Our data, reported below, suggest that scientists who receive NSF grants often obtain them by proposing projects with little or no agricultural import.

Scientists also are involved in a number of ways with a variety of nonfederal organizations. State governments provide large numbers of grants and contracts, as well as substantial numbers of consultancies, to agricultural scientists. To a lesser degree local governments also act in this role. Private foundations, private corporations, and commodity associations are important sources of extramural grants. In fact, private corporations are the single most important source of consulting for agricultural researchers. Publishers and international organizations are relatively unimportant.

Examination of the extraorganizational linkages of university and government scientists indicates that government scientists tend to receive few grants or contracts. This comes as no surprise, since grants are less important to government scientists than to university scientists in promotion evaluations and for funding of research. On the other hand, government scientists appear to consult to approximately the same degree as their university colleagues.

These summaries conceal significant differences among the disciplines. Differences are particularly acute with respect to grants, contracts, and consulting. In contrast, participation in advisory groups and review panels appears to be both rare and randomly distributed across disciplines.

Analysis of receipt of grants or contracts by discipline (Table 7.2) reveals that agricultural economists and social scientists receive grants and contracts mainly from USDA, foundations, and state governments. For agricultural engineers, the Department of the Interior, state governments, and private corporations provide fairly substantial grant support. Nevertheless, when compared to scientists in other fields, relatively few agricultural engineers receive grants or contracts from private corporations. This suggests that private corporations prefer to do their own engineering research, probably to insure that it does not enter the public domain. Agronomists, animal scientists, entomologists, and plant pathologists receive substantial numbers of grants from state governments, commodity associations, and private corporations. For example, 28% and 21% of animal scientists received grants and contracts, respectively, from private corporations during the last five years.

The basic sciences present quite a different pattern. More than a fifth of scientists in these fields have received grants from the National Science Foundation, and more than a quarter have received grants from the National Institutes of Health. In addition, over the last five years, 14% of basic scientists have been awarded foundation grants. A similar pattern holds for nutritionists, although private corporations play a larger role in funding nutrition research than they do in the basic sciences.

Extramural support for the environmental sciences comes largely in the form of contracts from USDA and grants from the Department of the Interior, foundations, state governments, and private corporations. Few environmental scientists receive grants from either NSF or NIH or from commodity associations. Among food scientists, foundation grants, commodity association grants, and private corporation grants and contracts are of approximately equal importance. Foresters rely most heavily for extramural support on USDA contracts, followed by significant numbers of NSF grants, state government grants and contracts, and private corporation grants and contracts. State governments, commodity associations, and private corporations are the major granting sources for horticulturists.

Table 7.2
Scientists' Grants or Contracts from Selected Organizations by
Discipline, 1974-1979

Discipline	USDA Contract [a]	Organization and Relationship (%)			
		Interior Grant [a]	NSF Grant[a]	NIH Grant[a]	Foundation Grant[a]
Agricultural Economics	32 [yz]	6 [y]	4 [xy]	1 [x]	12 [xy]
Agricultural Engineering	11 [w]	10 [y]	4 [xy]	1 [x]	6 [x]
Agronomy	16 [wx]	5 [y]	8 [xy]	1 [x]	11 [xy]
Animal Science	14 [wx]	2 [y]	8 [xy]	12 [y]	12 [xy]
Basic Sciences	11 [w]	2 [y]	21 [z]	26 [z]	14 [xy]
Entomology	17 [wxy]	0 [y]	13 [xyz]	5 [xy]	12 [xy]
Environmental Sciences	33 [z]	28 [z]	6 [xy]	4 [xy]	20 [y]
Food Science	8 [w]	2 [y]	8 [xy]	6 [xy]	16 [xy]
Forestry	40 [z]	1 [y]	13 [xyz]	0 [x]	5 [x]
Horticulture	12 [w]	1 [y]	2 [x]	0 [x]	13 [xy]
Nutrition	12 [wx]	2 [y]	15 [yz]	27 [z]	34 [z]
Plant Pathology	14 [wx]	0 [y]	9 [xyz]	5 [xy]	10 [xy]
Social Sciences	27 [xyz]	4 [y]	2 [x]	6 [xy]	13 [xy]

[a]Column differences, p < .01.

[b]Column differences, p < .05.

[v-z]Percentages within a column with different letters are significantly different, p < .01.

In general, contracts from USDA, foundation grants, and state government grants tend to be relatively evenly distributed across the disciplines. By contrast, grants from NSF and NIH tend to be concentrated in the basic sciences and nutrition, and grants from commodity associations and private corporations tend to be concentrated in those fields most closely linked to the concerns of various commodity groups and agribusinesses.

For the most part, grants appear to be distributed to scientists at various institutions without regard to institutional prestige or geographic region. However, there are several notable exceptions. National Science Foundation and National Institutes of Health grants tend to be concentrated in the most prestigious institutions and in those outside the South. Between a fifth and a quarter of the scientists at

TABLE 7.2 (cont.)

Discipline	State Government Grant[b]	Contract[a]	Commodity Grant[a]	Association Contract[b]	Private Grant[a]	Corporation Contract[a]
Agricultural Economics	12	11 ^{wxy}	4 ^{xy}	4	2 ^{v}	3 ^{y}
Agricultural Engineering	15	15 ^{xy}	7 ^{xyz}	8	10 ^{vwx}	8 ^{y}
Agronomy	17	9 ^{wx}	16 ^{yz}	6	19 ^{xyz}	12 ^{yz}
Animal Science	14	8 ^{wx}	10 ^{xyz}	5	28 ^{z}	21 ^{z}
Basic Sciences	9	3 ^{w}	7 ^{xyz}	2	10 ^{vwxy}	4 ^{y}
Entomology	16	7 ^{wx}	15 ^{yz}	4	24 ^{yz}	8 ^{yz}
Environmental Sciences	33	31 ^{z}	4 ^{xy}	0	15 ^{vwxyz}	9 ^{yz}
Food Science	8	6 ^{wx}	16 ^{z}	4	16 ^{vwxyz}	12 ^{yz}
Forestry	17	14 ^{wxy}	0 ^{x}	4	12 ^{wxy}	14 ^{yz}
Horticulture	20	7 ^{wx}	15 ^{yz}	10	16 ^{wxyz}	13 ^{yz}
Nutrition	10	5 ^{wx}	7 ^{xz}	2	20 ^{xyz}	15 ^{yz}
Plant Pathology	15	11 ^{wxy}	15 ^{yz}	9	15 ^{vwxyz}	9 ^{yz}
Social Sciences	19	21 ^{yz}	2 ^{x}	0	2 ^{vw}	4 ^{y}

[a] Column differences, $p < .01$.

[b] Column differences, $p < .05$.

$^{v-z}$ Percentages within a column with different letters are significantly different, $p < .01$.

the most prestigious institutions had received grants from one or both of those organizations during the last five years, whereas less than 10% of those in the least prestigious institutions had received such grants. Similarly, scientists outside the South are at least twice as likely to receive grants from those organizations as are those scientists in the southern region. Whether this reflects a consistent and deliberate bias on the part of the granting agencies or the quality of personnel in various regions and institutions is unclear (see Cole, Cole, and Simon, 1981; Cole, Rubin, and Cole, 1978; and the critical review of their work by Chubin, 1980a, 1980b).

Foundation grants also appear to be concentrated in the most prestigious institutions; over 20% of the scientists in those institutions report receiving foundation grants, but only 11% in the least prestigious institutions do. However, foundation grants do appear to be distributed equally across the various regions. Finally, private corporate grants tend to be distributed evenly across prestige groups but are concentrated in the southern region. Twenty-two percent of the scientists in the South have received corporation grants; in each of the other regions, 17% or fewer of the scientists have received such grants in the last five years ($p < 0.05$).

Grants and contracts from other agencies display no patterns of relationship with either region or prestige group. This suggests that most other types of grants and contracts are likely to be awarded to individuals at a particular institution chosen for the combination of competence and proximity to the area in which the commodity or problem to be researched is located.

Consulting is also unevenly distributed across the disciplines (Table 7.3), both by type of employer and by quantity. Among agricultural economists, entomologists, and plant pathologists, state governments were the major employers. Both state governments and private corporations commonly used agricultural engineers, agronomists, animal scientists, and food scientists. Those in the basic sciences frequently consulted for NSF; environmental scientists, for the Department of the Interior and state governments. Foresters consulted for a wide range of organizations, including state and local governments as well as private corporations. Commodity associations accounted for much consulting in horticulture, while nutritionists consulted equally for NIH and private corporations. Finally, many social scientists consulted for state and local governments.

Private corporations play a dominant role in providing consulting opportunities to agricultural scientists, particularly in food science (Table 7.3). One may well ask whether such high levels of private consulting can be justified given the mandate of the land-grant universities and USDA. The potential for conflicts of interest is certainly there (see Kenney et al., 1982).

Research Grants and Problem Choice

The sheer range and complexity of relationships between agricultural scientists and various sources of funding (e.g., grants, contracts, and consulting) make generalization about the impact of these activities on problem choice difficult. On the one hand, it might be argued that each of the various funding agencies would tend to draw to it scientists who shared that agency's values. Those scientists with a strong disciplinary orientation would be attracted toward certain govenment agencies such as the National Science Foundation, while

TABLE 7.3
Scientists Consulting for Selected Organizations by Discipline, 1974-1979[a]

Discipline	Interior 0[b]	NSF 0[c]	NIH 0[b]	Local Government 0[c]	State Government 0[b]	Commodity Association R[c]	Commodity Association 0[c]	Private Corporation R[c]	Private Corporation 0
Agricultural Economics	4 xy	3 y	0 x	7 xyz	17 xyz	5 yz	4 yz	7 xy	19
Agricultural Engineering	2 xy	1 y	0 x	8 xyz	15 xyz	2 y	7 yz	11 xy	29
Agronomy	5 xy	4 yz	1 xy	9 xyz	10 wxy	4 yz	5 yz	11 xy	20
Animal Science	3 xy	6 yz	5 xyz	4 xy	7 wxy	5 yz	6 yz	9 xy	18
Basic Sciences	3 xy	14 z	7 yz	3 xy	5 wx	1 y	6 yz	8 xy	16
Entomology	2 xy	7 yz	3 xyz	4 xy	15 xyz	5 yz	3 yz	4 xy	15
Environmental Sciences	20 z	6 yz	0 x	9 xyz	26 z	0 y	0 yz	4 xy	24
Food Science	2 xw	6 yz	2 xy	6 xy	12 wxy	8 y	8 yz	24 z	24
Forestry	9 y	6 yz	0 x	13 yz	16 xyz	4 yz	3 y	14 yz	22
Horticulture	2 xy	3 y	2 xy	7 xy	9 wxy	12 z	12 z	9 xy	14
Nutrition	0 x	5 yz	10 z	0 x	0 w	0 y	0 y	10 xy	12
Plant Pathology	1 xy	8 yz	4 xyz	4 xy	13 wxyz	1 yz	7 yz	4 xy	20
Social Sciences	9 y	6 yz	4 xyz	17 z	19 yz	0 y	2 y	2 x	11

[a] 0 = occasional consultant; R = regular consultant.

[b] For column differences, p < .001.

[c] p < .01

w-z Percentages within a given column with different letters are significantly different, p < .01.

those with strong client orientations would be attracted to commodity associations and private corporations. An alternative interpretation would be that the funding agency directly or indirectly influenced the direction and scope of research. That is, in a situation in which research money was limited, scientists would direct their research toward those problems of interest to funding agencies. A final possible interpretation would be that scientists use research grants to "bootleg" (Greenberg, 1966) their own interests. In such cases, we should find scientists providing the appropriate "buzzwords" for granting agencies while quietly going about what they had planned to do anyway.

Of course, to some extent these three interpretations are not mutually exclusive. A given scientist may enter into a particular relationship with an external funding source in order to further his or her own professional or personal aims. At the same time, that individual may satisfy the research aims of the funding agency. And, finally, that same scientist may begin to identify the approach, goals, and interests of the funding source as his or her own.

In order to address this issue, it was necessary to assemble statistical data relating the receipt of grants, contracts, or consulting assignments to criteria for problem choice. The statistical information was supplemented with unsolicited comments from scientists that illuminate more clearly how extraorganizational influences are related to problem choice.

To simplify the task of examining the statistical information, we focus upon the receipt of research grants from four major sources: two public basic research institutions (Table 7.4) and two major sources of private funds (Table 7.5). Data on research contracts, consulting, and other types of relationships reveal a similar, though weaker, set of relationships with problem choice.

Recipients of grants from the National Science Foundation and the National Institutes of Health differ significantly from their colleagues on eleven out of twenty-one criteria for problem choice. These differences form a relatively straightforward pattern. NSF and NIH grant recipients tend to weight only three items more highly than their colleagues: the contribution to scientific theory, enjoyment, and scientific curiosity. In short, grant recipients are more concerned about the disciplinary contribution that their research may make than are their colleagues. In addition, grant recipients see each of the other eight items listed in the table as significantly *less* important in decisions about problem choice. NSF and NIH grant recipients give less weight to clients, either broadly or narrowly defined, as well as to the priorities of their own research organization, than do their colleagues who have not received such grants.

TABLE 7.4
Selected Criteria for Problem Choice by Receipt of Grants from the National Science Foundation and the National Institutes of Health[a]

| | Organization | | | |
| | NSF Grant | | NIH Grant | |
Criterion	Yes	No	Yes	No
Enjoyment	6.1[c]	5.8	6.2[c]	5.8
Scientific curiosity	5.8[b]	5.1	5.9[b]	5.1
Contribution to scientific theory	5.2[b]	4.2	5.7[b]	4.2
Importance to society	5.2[b]	5.6	5.3[c]	5.6
Client needs assessment	3.7[b]	5.0	3.1[b]	5.0
Organizational priorities	3.5[b]	4.4	3.4[b]	4.4
Clientele demands	3.2[b]	4.2	2.5[b]	4.3
Marketability	2.8[b]	3.8	2.2[b]	3.9
Extension feedback	2.7[b]	3.6	2.3[b]	3.6
Publication in bulletins	2.5[b]	3.8	2.0[b]	3.8
Publication in farm journals	2.2[b]	3.1	1.8[b]	3.1

[a]Mean scores based on a seven point scale (1 = not important, 7 = very important).

[b]Row pair differences, p < .001.

[c]Row pair differences, p < .01.

TABLE 7.5
Selected Criteria for Problem Choice by Receipt of Grants from Private Corporations and Commodity Associations[a]

| | Organization | | | |
| | Private Corporation Grant | | Commodity Association Grant | |
Criterion	Yes	No	Yes	No
Availability of facilities	5.6[b]	5.2	5.6[b]	5.2
Creation of new methods	5.4[b]	5.0	5.2[b]	5.0
Marketability	3.9[b]	3.7	4.2[b]	3.7
Extension feedback	3.9[b]	3.4	3.9[c]	3.4
Client needs assessment	5.3[b]	4.8	5.6[b]	4.8
Clientele demands	4.5[b]	4.1	4.7[b]	4.1
Publication in farm journals	3.3[b]	2.9	3.4[b]	2.9

[a]Mean scores based on seven point scale (1 = not important, 7 = very important).

[b]Row pair differences, p < .001.

[c]Row pair differences, p < .01.

Recipients of grants from private corporations and commodity associations reflect a similar, though inverse, relationship with regard to clients. Grant recipients from these organizations appear significantly more concerned with clients than their colleagues who have not received such grants. (It should be noted that "nonrecipients" in the context below include all scientists not reporting receipt of a grant, whether or not they applied for one.) Recipients of corporation grants tend to be interested in the creation of new methods, materials, and devices to a greater extent than their nonrecipient colleagues. In contrast, recipients of commodity association grants tend to be more concerned about the marketability of the final product than their nonrecipient colleagues. This difference clearly reflects the difference between private corporations interested in developing new products and commodity associations generally organized around and interested in the marketing of existing products.

Of equal importance are those items upon which grant recipients and nonrecipients do not significantly differ. For example, funding is no more important as a criterion in problem choice to grant recipients than to nonrecipients. This suggests that those who have received grants may no longer see funding as a criterion or constraint upon problem choice. Nor are researchers who received grants more concerned with hot topics or the evaluation, approval, or credibility of their colleagues doing similar research.

How do we interpret these results? Is it that scientists with strong disciplinary orientations go to the National Science Foundation, while those with strong client orientations go to private corporations and commodity associations? Or are scientists' problems, in part, redefined by the funding agencies? The final possibility, that scientists "bootleg" their research, appears to be clearly contradicted by this statistical evidence; if bootlegging were a major factor, then relationships between criteria for problem choice and the receipt of grants would be relatively minor or nonexistent.

With these two possibilities in mind, let us examine what scientists in various disciplines who have received grants report:

Funding, the availability of it, and who is the granting agency determines the kind of research that will be done. [An entomologist]

Funding appears to dictate activities and opinions. As an example, if funding is pesticide oriented—and comes from industry—all pesticides are desirable and if funding is environmentally oriented, no pesticide is good. The truth lies somewhere between the two; our funding sources polarize activities. [A horticulturist]

Personally I have usually been fortunate in obtaining funds to work in areas that were interesting to me and which I felt were important. But, I have seen many good scientists get used by money and end-up very disappointed and discouraged with research. [A soil scientist]

Agricultural science is entirely too dependent for success on grants-in-aid from big money industries unrelated to agriculture. [A plant pathologist]

The general decline in hard funds forces too many to seek (and we have been generally successful) grants (NSF, NIH, commodity, etc.). This in many cases places the need for funding in a higher priority than where research needs are greatest—in short, it controls research directions too much. [An entomologist]

The single most important factor in my decision as to the direction of my research is available support. Grants are essential to the ongoing of my research program. [An engineer]

Perhaps, pressure, policy, and directions of influence by funding and funding agencies is not given enough consideration. These are major influences. In fact, they are turning some agricultural research workers, departments, and programs away from former agricultural orientations. [A nutritionist]

Of course, short of following around hundreds of scientists for an extended period of time, no definitive description of the relationship between grant receipt and problem choice can be developed. In sum, it is likely that in almost all cases: (1) scientists choose potential granting organizations according to their interests and the probability of receipt of funds, *and* (2) those granting organizations have an influence on the scientists.

The research support for this book is an example of the relationship between funding and research. Initially, we sought and received experiment station funds. Then, we attempted to secure extramural funds from the National Science Foundation and the Ford Foundation. (The latter attempt was successful; the former was not.) Clearly, in deciding what organizations to approach, we excluded commodity groups and private corporations, as we knew in advance that they would not be likely to fund our work. Moreover, our proposal to NSF focused upon the disciplinary aspects of our research rather than the policy issues, in a deliberate attempt to describe the proposed research in a manner that would be favorably received by the reviewers. For the Ford Foundation, our proposal emphasized our mutual interests in policy questions. Had we received the NSF grant, we probably would have focused to a greater extent upon the disciplinary aspects of our work and written a more sociological account.

Our grant from the Ford Foundation has influenced us in both direct and subtle ways, and probably in many ways of which we are not fully aware. Using foundation funds, we have held workshops involving science policy makers, administrators, research directors of private corporations, and small farmers. We have become friendly with foundation staff, and have been urged to pursue certain lines of research in furtherance of certain ends. In short, we have both been selective in our choice of funding sources, and been influenced by them.

Nevertheless, the comments made by scientists are clear. Funding and related activities appear all too often to dictate the direction and, occasionally, even the output of agricultural research. As one experiment station associate director warned some twenty-five years ago: "'Captive' is a good description of the situation. The donors of these grants are attempting to progress at the expense of the captive scientists, and the captive scientists are willing to become captive because they can grind out more publications by a greater number of graduate students and thus progress to more lucrative positions . . . " (Hamilton, 1954:74). Though the author admittedly overstated the case, the message remains a valid one. It is all too easy to accept the funding agency's agenda as one's own without further inquiry (Feigenbaum, 1980).

Conclusions

The current fiscal squeeze can only serve to magnify this problem by increasing the pressures on scientists to seek private grants and consultancies and on administrators to accept all that are proffered (Hadwiger, 1982). Public agencies, at least in principle, make grants with public purpose in mind. Private agencies need not have such a continuity of interest with public agricultural research.

Time spent searching for grants is time that might otherwise be used for research. If grants are both plentiful and relevant to the broadly defined goals of the USDA and the land-grant universities, then such searching can have successful outcomes. However, as funds become scarce, the probability of any given scientist obtaining a grant diminishes, and the costs surrounding grant-seeking increase.

Finally, the public agricultural scientist works within an environment that is, in a very real sense, a political one. An individual scientist may work on a given topic and remain seemingly oblivious to politics, but the very fact that funds are available to pursue that topic is, in part, the result of overtly political actions on the part of both elected officials and various interest groups. These actions, as well as the actions of scientists, determine who benefits from agricultural research. It is this issue that we address in the following chapter.

8
Research for Whom?

Who benefits from agricultural research? On the surface this would appear a relatively simple question. However, the debate of the last decade attests to the contrary. Some suggest that only agribusiness benefits from agricultural research, whereas others argue that farmers are the major beneficiaries. Still others argue that the general public has benefited most. Even official representatives for the agricultural sciences have vacillated in arguing whether farmers or consumers have been the major beneficiaries of agricultural research.

Scientists' assumptions about the beneficiaries of their research reflect their family backgrounds, communication patterns, disciplines, organizations, and extraorganizational ties, and in turn affect their choice of research problems. The influence of perceptions about beneficiaries on problem choice makes it important to review also the relatively sparse empirical literature on beneficiaries of agricultural research.

Scientists' Perceptions of Beneficiaries

To assess scientists' perceptions of the beneficiaries of their research, we asked them the following questions: "Apart from your discipline, do you believe that your research and publishing over the past five years has already or will directly or indirectly benefit any of the following? In your opinion who should your research benefit?" Scientists were provided with a list of potential beneficiaries and asked to rate each on a scale ranging from 1 (not at all) to 5 (a great deal). The mean ratings for each group are shown in Figure 8.1. As the figure reveals, large farmers and the general public are the most important perceived beneficiaries, followed by other scientific disciplines, small farmers, and agribusiness. Rural residents, local, state, and federal agencies, and foreign groups, institutions, or governments are somewhat less important. (Farm workers were inadvertently omitted from

FIGURE 8.1
Perceived Actual and Preferred Beneficiaries of Agricultural
Research (Mean Scores).

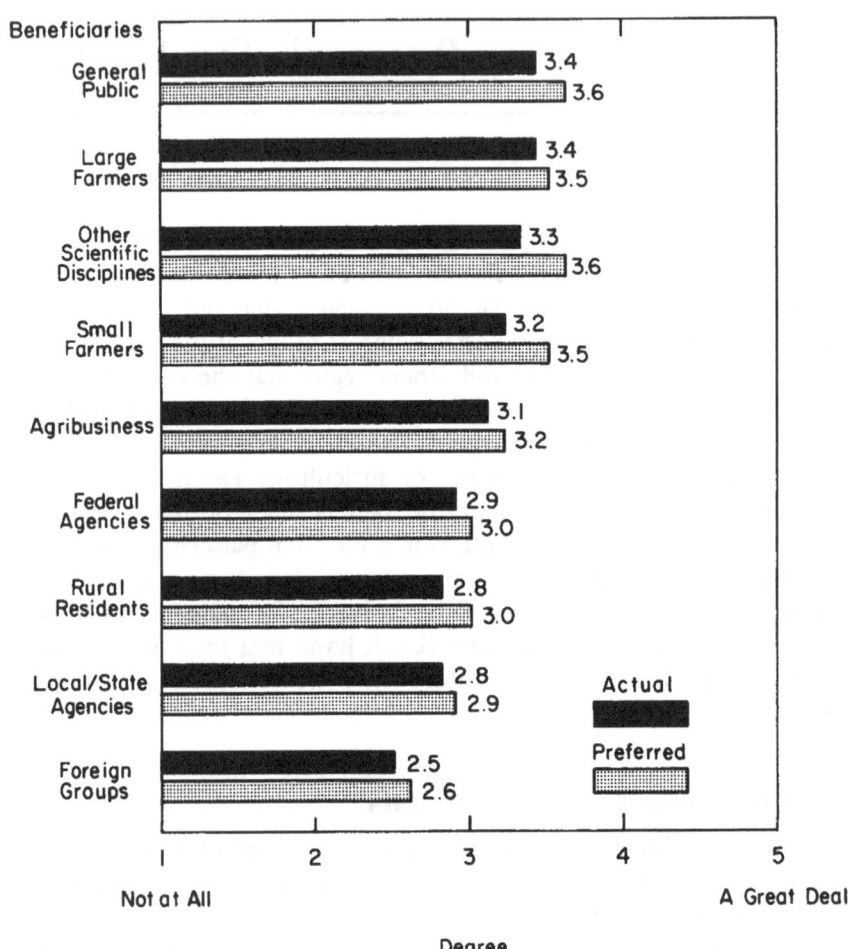

the list; see the Methodological Appendix.) When asked who *should*
benefit from their research, scientists rated each group somewhat
higher. Of particular note are the scientists' views that, ideally, small
farmers should benefit as much as large farmers from agricultural
research, although they are currently perceived to be at a slight actual
disadvantage.

The relatively minimal differentiation among beneficiaries is surprising (Figure 8.1). Mean scores vary only 1 point on the 5 point

TABLE 8.1
Beneficiaries of Research by Father's Occupation [a]

| | Father's Occupation | |
Beneficiary	Farmer	Other
Other disciplines	3.3 [b]	3.1
Small farmers	3.3 [b]	3.1
Large farmers	3.6 [b]	3.2
Agribusiness	3.4 [b]	3.0
Rural residents	3.0 [b]	2.7
General public	3.5	3.3
State and Local Government	2.7	2.9
Federal Government	2.8	2.9
Foreign	2.4	2.5

[a]Mean scores on a five point scale (1 = not at all,
5 = a great deal).

[b]Row pair differences, $p < .001$.

scale. However, as in the case of criteria for problem choice, mean
scores conceal significant differences within the scientific community.
These differences appear attributable to factors ranging from the
background of the scientists to the extraorganizational influences upon
them.

Farm Background

Scientists whose fathers were farmers see their research as significantly more likely to benefit the farming and rural communities than
those scientists whose fathers had nonfarm occupations (Table 8.1).
In addition, scientists with farm backgrounds are most likely to see
large farmers as the major beneficiaries of their work. On the other
hand, there is no difference between the two groups in terms of
perceived benefits for other disciplines, the general public, or various
governments. Future entrants into the agricultural science system are
less and less likely to be drawn from farm backgrounds. Therefore,
farm and rural populations may become less important as target
populations for agricultural research.

Communication Frequency

The frequency with which scientists communicate about their research also appears to affect who they perceive as beneficiaries (Table
8.2). Scientists who communicate daily with colleagues within their

TABLE 8.2
Research Beneficiaries by Scientists' Communication with System Participants[a]

System Participants	Beneficiary								
	Other Disciplines	Small Farmers	Large Farmers	Agri-business	Rural Residents	General Public	Local/St. Government	Federal Agencies	Foreign Groups
Scientists within-in department									
Rarely	3.3	3.0	3.1	3.0	2.6	3.4	2.7	2.8[b]	2.3[b]
Daily	3.5	3.2	3.5	3.2	2.9	3.5	3.0	3.1	2.7
Other Agricultural scientists									
Rarely	3.2[b]	2.7[b]	2.8[b]	2.7[b]	2.6[b]	3.3	2.7	2.8	2.2[b]
Daily	3.5	3.5	4.1	3.8	3.3	3.6	2.6	2.8	2.7
Nonagricultural scientists									
Rarely	3.1[b]	3.4[b]	3.6[b]	3.3	2.9	3.3	2.7	2.7[b]	2.4
Daily	3.9	2.8	2.9	3.4	2.7	3.4	3.1	3.2	2.8
Administrators									
Rarely	3.3	3.0[b]	3.2[b]	3.0[b]	2.6[b]	3.3	2.7	2.7[b]	2.4[b]
Daily	3.4	3.6	3.9	3.4	3.4	3.6	2.8	3.2	2.9
Clients									
Rarely	3.4	2.8[b]	2.9[b]	2.7[b]	2.5[b]	3.1[b]	2.5[b]	2.7	2.3
Daily	3.6	3.9	4.2	3.8	3.2	3.8	3.1	3.3	2.8
Funding agencies									
Rarely	3.3	3.1	3.3	3.1	2.7[b]	3.3[b]	2.6	2.7[b]	2.4
Daily	4.0	1.5	2.5	2.5	1.5	3.5	2.0	3.5	1.5
Extension staff									
Rarely	3.4	2.7[b]	2.8[b]	2.7[b]	2.5[b]	3.3	2.7	2.9	2.3
Daily	3.2	3.9	4.2	3.8	3.5	3.6	2.8	2.7	2.6

[a]Mean scores based upon five point scale (1 = not at all, 5 = a great deal).

[b]Column pair differences, $p < .01$.

own department are more likely to see other disciplines, large farmers, and federal agencies as beneficiaries of their research than their colleagues who communicate only rarely with others within their own department. Those who communicate frequently with agricultural scientists outside their departments appear more likely to see other disciplines and all farm and rural groups as more likely to benefit from their research. However, while those who communicate rarely see little difference among large farmers, small farmers, and agribusiness as beneficiaries, those who communicate daily with agricultural scientists outside their departments appear to have a somewhat stronger large farm and agribusiness orientation.

Scientists who communicate daily with nonagricultural scientists appear quite likely to see other disciplines as the primary beneficiaries of their research. They see both small and large farmers as benefiting somewhat less than do their colleagues who communicate only rarely with nonagricultural scientists. However, those who communicate daily with nonagricultural scientists also see agribusiness and federal agencies as more likely to benefit from their research than do those who communicated infrequently.

Those who communicate more frequently with administrators are also more likely to see all groups except other disciplines and the general public as beneficiaries of their research. A similar pattern holds for communication with clients; frequent communication is positively related to the perception of benefits for all except those in other disciplines. In the latter case, no relationship exists.

In the case of communication with funding agencies, those who communicate frequently are more likely to see the general public, other disciplines, and federal agencies as beneficiaries. On the other hand, they see large farmers, agribusiness, rural residents, and local and state government as less likely to benefit from their research. Finally, and not surprisingly, frequent communication with extension staff is positively related to the perception of benefits to all except those in other disciplines and those who work for federal agencies.

Discipline

Scientists' perceptions of beneficiaries of their research also differ markedly by discipline (Table 8.3). Among agricultural economists, the general public is seen as the most likely beneficiary of research, and other disciplines as the least likely. Large farmers are seen as the most important beneficiaries of research in agricultural engineering, agronomy, animal science, entomology, horticulture, and plant pathology. Engineers and agronomists classify rural residents, various levels of government, and foreign groups at the bottom of their list

TABLE 8.3
Research Beneficiaries by Discipline[a]

Discipline	Other Disciplines[b]	Small Farmers[b]	Large Farmers[b]	Agri-business[b]	Rural Residents[b]	General Public	Local/St. Government[b]	Federal Agencies[b]	Foreign Groups
Agricultural Economics	2.7[w]	3.1[xy]	3.1[xy]	3.3[z]	3.0[y]	3.5	3.4[xyz]	3.3[xy]	2.2
Agricultural Engineering	3.3[xyz]	3.3[yz]	3.6[yz]	3.4[z]	3.0[y]	3.5	3.0[wx]	3.1[wxy]	2.6
Agronomy	3.3[xyz]	3.6[yz]	3.8[z]	3.4[z]	2.9[y]	3.3	2.9[vwx]	2.9[wxy]	2.6
Animal Science	3.3[xyz]	3.5[yz]	4.0[z]	3.6[z]	3.0[y]	3.3	2.3[tu]	2.3[tu]	2.6
Basic Sciences	3.6[yz]	2.5[wx]	2.7[wx]	2.6[y]	2.1[w]	3.2	2.0[t]	2.4[tuv]	2.2
Entomology	3.4[xyz]	3.4[yz]	3.6[yz]	3.3[z]	2.8[xy]	3.3	2.8[uvw]	3.0[wxy]	2.6
Environmental Sciences	3.6[yz]	2.7[wx]	2.7[wx]	2.4[xy]	2.9[y]	3.5	3.9[z]	3.9[z]	2.7
Food Science	3.5[yz]	2.5[w]	2.7[wx]	3.6[z]	2.4[wx]	3.7	2.8[uvw]	2.8[uvwx]	2.5
Forestry	3.1[wxy]	2.4[w]	2.4[w]	1.9[x]	2.9[xy]	3.2	3.3[wxy]	3.4[yz]	2.3
Horticulture	3.0[wx]	3.9[z]	4.0[z]	3.5[z]	3.0[y]	3.6	2.5[tuv]	2.2[t]	2.3
Nutrition	3.7[z]	2.4[w]	2.4[w]	2.6[y]	2.6[wxy]	3.6	2.6[tuv]	2.6[wxxy]	2.5
Plant Pathology	3.4[xyz]	3.7[z]	4.0[z]	3.4[z]	2.7[wxy]	3.2	2.6[uw]	2.9[tuvw]	2.7
Social Sciences	3.4[xyz]	2.4[w]	2.3[w]	2.1[xv]	3.9[z]	3.9	3.7[yz]	3.4[yz]	2.3

[a]Mean scores based upon five point scale (1 = not at all, 5 = a great deal). [b]Column differences, p < .001.

[t-z] Means within a column with different letters are significantly different, p < .01.

of beneficiaries. State, local, and federal government, and foreign groups are also seen as the least important beneficiaries of research in animal science, horticulture, and plant pathology. Entomologists see rural residents, local and state governments, and foreign groups as least likely to benefit from their research.

Interestingly, basic scientists perceive other disciplines as the most important beneficiaries of their research. Given the very low level of reported influence by basic scientists on agricultural scientists' research decisions (Chapter 4), other agricultural science disciplines probably are not among the actual beneficiaries of their research. State and local government are the least important beneficiaries perceived by researchers in the basic sciences. Among nutritionists, other disciplines also rank as the most important beneficiaries; here, however, both small and large farmers are the negligible beneficiaries. Environmental scientists rank state, local, and federal agencies at the top of their list of beneficiaries, and agribusiness at the bottom.

Food scientists, like their colleagues in social sciences, see the general public as the major beneficiary of their research and rural residents as the least important beneficiaries. Foresters, reflecting the strong ties between forest research and the national forests, see federal agencies as their major beneficiaries, and agribusiness as least likely to benefit from their research. On the surface, this finding appears counterfactual; it is clear that forest and wood products firms have benefited substantially from forest research. On the other hand, just as many foresters do not consider themselves to be "agricultural scientists" (see the Methodological Appendix), many may also not consider such firms to be included under the rubric "agribusiness." Finally, social scientists see rural residents and the general public as the most important beneficiaries of their research, and agribusiness as the least important beneficiary.

These data may also be viewed from the perspective of the potential beneficiaries. For example, it appears that other disciplines have most to gain from nutritionists and basic scientists. Small farmers are perceived to benefit most from horticultural research and least from research in forestry, nutrition, and the social sciences. Large farmers appear to benefit equally from research in horticulture, animal science, and plant pathology, but like small farmers, they are felt to benefit little from research in forestry, nutrition, and social sciences. Agribusiness is seen as benefiting most from research in agricultural engineering, animal science, and food science, and least from research in forestry. Far and away, social scientists are most likely to perceive their research as benefiting rural residents, while basic scientists perceive themselves as contributing the least to this group. Social scientists most often perceive themselves as benefactors of the general public, while researchers in the basic sciences and plant pathology least often see themselves in this way. Researchers in the environmental sciences perceive their research as most beneficial for local, state, and federal agencies. Scientists in the basic sciences see themselves as least important for local and state government, while horticulturists perceive their contribution as least important for federal agencies.

Most strikingly, large farmers are seen as the most important beneficiaries of research by members of six disciplines. In contrast, no discipline perceives small farmers as its most important beneficiary. In only one discipline do scientists see small farmers as more likely to benefit from their research than large farmers. Moreover, nutritionists see their research as more likely to benefit agribusiness than farmers of any size. In sum, there appears to be a consistent, though admittedly small, bias in perceptions of the beneficiaries of research toward larger farmers and agribusiness and away from smaller farmers.

Organization

The institutional context within which scientists work appears not to affect their perceptions of beneficiaries. The only exception is a slightly higher score given to foreign groups by government scientists (university = 2.4, government = 2.7, $p < 0.005$). On the other hand, there are significant differences across regions and prestige categories.

Perceived beneficiaries by experiment station region are depicted in Table 8.4. Scientists in the southern region appear more likely to see small farmers, large farmers, agribusiness, and rural residents as beneficiaries than their colleagues in other regions. Scientists in the western region are especially likely to see their research as benefiting local or state government or federal agencies. These data appear to be consistent with those reported in Chapter 6, which emphasized the client-oriented approach of scientists in the southern region.

Scientists' perceptions of beneficiaries also vary with institutional prestige (Table 8.5). Scientists in the most prestigious institutions are less likely to see farmers, agribusiness, or rural residents as beneficiaries of their research. However, no difference across prestige groups exists in terms of perceived benefits to other disciplines, the general public, or various levels of government. Here again, these results confirm those reported in Chapter 6 concerning the role of institutional prestige.

Grant Receipt

Given the strong influence of grants in problem choice, it appeared probable to us that grants would also play a role in scientists' perceptions

TABLE 8.4
Research Beneficiaries by Region[a]

	Region			
Beneficiary	Northeast	North Central	Southern	Western
Other disciplines	3.3	3.3	3.3	3.4
Small farmers	2.9[b]	3.0	3.4	3.2
Large farmers	3.1[b]	3.2	3.7	3.4
Agribusiness	2.9[b]	3.0	3.4	3.1
Rural residents	2.7	2.9	3.0	2.7
General public	3.3	3.5	3.3	3.4
Local or state government	2.8[b]	2.7	2.7	3.1
Federal agencies	2.7	2.7	2.6	3.0
Foreign groups	2.3	2.3	2.5	2.5

[a]Mean scores based upon five point scale (1 = not at all, 5 = a great deal).

[b]Row differences, p < .01.

TABLE 8.5
Research Beneficiaries by Prestige of Institution[a]

	Prestige	
Beneficiary	Least Prestigious	Most Prestigious
Other disciplines	3.3	3.3
Small farmers	3.3[b]	2.7
Large farmers	3.5[b]	2.9
Agribusiness	3.2[b]	2.8
Rural residents	2.9	2.6
General public	3.4	3.3
Local or state government	2.9	2.8
Federal agencies	2.8	2.7
Foreign groups	2.4	2.3

[a]Mean scores based upon five point scale (1 = not at all, 5 = a great deal).

of beneficiaries. The data appear to support this contention (Table 8.6). Scientists who have received grants from the National Science Foundation see their research as significantly more likely to benefit other disciplines and less likely to benefit nearly all other groups. A similar pattern holds for those scientists who have received grants from the National Institutes of Health. However, though NIH grant recipients do not differ from nonrecipients in their ranking of the general public as a beneficiary of their research, those who have received NSF grants see the general public as less likely to benefit than do nonrecipients. The recipients of Interior Department grants see farmers as less likely to benefit and state, local, and federal agencies as more likely to benefit from their research than do nonrecipients.

Recipients of commodity association and private corporation grants tend to perceive beneficiaries quite differently from their colleagues who have received grants from federal agencies. Specifically, private sector grant recipients see farmers, both small and large, and agribusiness as more likely to benefit from their research than do nonrecipients. Once again, however, the slight bias in favor of large farmers is greater among grant recipients than among nonrecipients. We found little or no difference between recipients' and nonrecipients' perceptions of other groups of beneficiaries. These findings are fairly consistent with those relating the receipt of grants to criteria for problem choice.

In summary, scientists tend to see a rather wide variety of groups as benefiting from their research. Scientists who are linked to the farm and agricultural communities are more likely to see farmers as beneficiaries of their research than are those without such linkages. On the other hand, not having those linkages does not necessarily imply that one perceives the general public as a more important beneficiary of research. Indeed, among scientists from nonfarm backgrounds, the scientific community is a more likely beneficiary. Finally, the more contact a scientist has with the farm and agricultural sector, the greater his or her tendency to perceive large farmers rather than small farmers as potential beneficiaries of research. Such a bias, however small, would if consistent lead to a cumulative advantage over time for the large farmers.

Beneficiaries and Problem Choice

All research agendas are not equally likely to benefit all groups of beneficiaries. Most scientists are aware of this when they choose their problems. Thus, although they may not actively investigate the probable consequences of their research, their image of who benefits may well

TABLE 8.6
Grantees' and Nongrantees' Perceptions of Beneficiaries of Their Research, Selected Granting Agencies[a]

Funding Source	Beneficiary								
	Other Disciplines	Small Farmers	Large Farmers	Agri-business	Rural Residents	General Public	Local/St. Government	Federal Agencies	Foreign Groups
National Science Foundation Grant									
Yes	3.7[b]	2.4	2.5[b]	2.4[b]	2.2[b]	2.9[b]	2.5	2.7	2.4
No	3.3	3.2	3.4	3.2	2.9	3.4	2.8	2.9	2.5
National Institutes of Health Grant									
Yes	3.9[b]	2.3[b]	2.4[b]	2.4[b]	2.2[b]	3.3	2.2[b]	2.6	2.4
No	3.3	3.2	3.4	3.2	2.9	3.3	2.9	2.9	2.5
Department of Interior Grant									
Yes	3.5	2.4[b]	2.8[b]	2.4[b]	2.9	3.5	3.7[b]	3.6[b]	2.3
No	3.3	3.2	3.4	3.2	2.8	3.4	2.8	2.8	2.5
Commodity Association Grant									
Yes	3.3	3.7[b]	4.1[b]	3.8[b]	3.0	3.5	2.7	2.7	2.6
No	3.3	3.1	3.3	3.0	2.8	3.4	2.8	2.9	2.4
Private Corporation Grant									
Yes	3.4	3.7[b]	4.0[b]	3.6[b]	3.0	3.5	2.8	2.7	2.6
No	3.3	3.1	3.2	3.0	2.8	3.3	2.8	2.9	2.4

[a]Mean scores based upon five point scale (1 = not at all, 5 = a great deal).

[b]Column pair differences, p < .01.

affect the relative weight they give to various criteria for problem choice.

At the same time, having chosen particular criteria for problem choice, it is likely that scientists will begin to see certain groups as beneficiaries and others as unlikely to benefit. In short, relations between perceptions of beneficiaries and criteria for problem choice are both complex and multidimensional. No causal relation—in even the loosest sense—between the two groups of variables can be definitively stated, certainly not with the crude survey data that we report here. Yet, there is likely to be a mutual interdependence. Therefore, we report the correlations between selected criteria for problem choice and perceptions of selected beneficiaries of research (Table 8.7). We have grouped the correlations in such a way as to display a pattern in the relationships.

The first group of criteria listed in the table—contribution to science, enjoyment, and scientific curiosity—are all positively related to perceptions of other scientific disciplines as beneficiaries. On the other hand, these criteria are all negatively related or unrelated to perceptions of benefit to the other groups listed. In contrast, the criteria in the second group, all of which have stronger associations with applied research, are positively correlated with each of the possible beneficiary groups, *except* other disciplines, where the correlations are weak, but negative. The correlations are strongest between this group of criteria and perceived beneficiaries among the traditional clientele of agricultural research (farmers and agribusiness).

The last group of criteria in the table are positively correlated with all beneficiary groups. Of particular interest are the relatively weak correlations between "organizational priorities" and perceptions of either other disciplines or the general public as beneficiaries.

Of equal interest are several items not listed in the table. For example, "publication probability in professional journals" is significantly correlated ($r = 0.14$) only with seeing "other disciplines" as beneficiaries of one's research. Similar patterns were found for "evaluation of research by scientists in your field" and "credibility of other investigators doing similar research."

These correlations underscore once again the apparent tension between disciplinary and practical concerns that we found throughout our study. Of course, most agricultural scientists do not conduct empirical studies of the beneficiaries of their research. Instead, like scientists in other fields, they assume that certain groups and individuals will or will not benefit from their research. It therefore behooves us to look at the somewhat scanty empirical literature on the beneficiaries of agricultural research.

Empirical Studies of Beneficiaries

Most studies of the beneficiaries of agricultural research tend to focus upon one or two groups. Furthermore, available studies appear to be divided into two types: (1) case studies of particular products of agricultural research, and (2) quantitative studies of spatial entities (counties, states, etc.) that generally employ secondary data. Unfortunately, it is often difficult, if not impossible, to link the detailed case studies to those that take an aerial view.

Farmers

One of the earliest discussions of the beneficiaries of agricultural research is to be found in an article by Henry Ernest Erdman that appeared in the *Journal of Farm Economics* in 1929. Erdman was concerned about the impact of agricultural science and technology on farming. He concluded that advantage accrued to farmers who most rapidly adopted new technologies: "Now all of this means that the individual who adopts improvements increases his profits, *provided not enough other farmers do likewise to lower price*" (1929:30; emphasis in original). In fact, research that increases all farmers' productivity floods the market, thereby reducing prices and benefiting only the consumer. Erdman also saw a bias in new technology toward reducing the need for so many farmers, and made a number of suggestions for ameliorating the problem. His solutions included encouraging farmers' sons to enter other occupations, rural industrialization, and a policy of government purchase of marginal land (1929).

A later study by Earl Heady discussed the distributional impact of agricultural research:

> New techniques may transfer income and wealth between individuals regardless of whether total net farm income is increased or decreased. The transfer may be of an intra-industry nature in the sense that it increases the income of some people in agriculture at the expense of other people within the industry. Income will be transferred when the techniques for one commodity or geographic region are improved beyond that which applies for a competing commodity or region (1949:305).

More recently, a USDA–sponsored study reached much the same

TABLE 8.7
Correlations Between Selected Criteria for Problem Choice and Selected Beneficiaries of One's Research[a]

Criterion	Other Disciplines	Beneficiary				
		Small Farmers	Large Farmers	Agribusiness	Rural Residents	General Public
Group I						
Contribution to scientific theory	.36	-.11	-.11	-.12	-.09	.00
Enjoyment	.20	-.02	-.02	.01	.01	-.01
Scientific curiosity	.21	-.10	-.09	-.08	-.04	.02
Group II						
Marketability	-.03	.31	.33	.35	.27	.19
Publication in farm/industry journals	-.05	.35	.35	.30	.27	.12
Publication in bulletins	-.08	.30	.29	.25	.29	.11
Clientele demands	-.11	.27	.31	.27	.24	.11
Extension feedback	-.07	.35	.35	.30	.25	.07
Assessment of client needs	-.10	.36	.40	.34	.30	.21
Group III						
Importance to society	.10	.24	.24	.21	.28	.27
Organizational priorities	.02	.21	.22	.22	.13	.07

[a]Pearsonian correlation coefficients (r). Values range from -1.0 (strong negative relationship) to 1.0 (strong positive relationship). A value of 0.0 is interpreted as the complete absence of any relation between the two variables. As the data is ordinal and the coefficients assume interval data, reported results are conservative. If r > ± .09, then p < .001.

conclusion about the relationships among technology, farm size, and shifts in the location of production (Babb, 1979).

In part, the bias toward large farms noted in these and other studies can be traced to the greater capital requirements frequently associated with agricultural innovations. As a 1978 GAO report put it, "The entire agriculture system—from agriculture schools and colleges to the research community—also tries to find ways to make farmers extensively more productive through capital-intensive farming and mechanization" (U.S. General Accounting Office, 1978:25). As small farmers frequently have greater difficulty in obtaining access to capital and pay somewhat higher rates for it, highly capitalized enterprises tend to be found only on large farms (Kerr, 1980).

Initially, therefore, the early adopters were the key beneficiaries of agricultural research. Adoption was likely to lead to an increase in farm size and a resource base that made possible future adoption. Consequently, the research system contributed to the emergence of client groups (large farmers) who were more likely to be served with each subsequent innovation or technology.

Recently, agricultural mechanization has been singled out as particularly important in this regard. Without question, agricultural mechanization has reduced the drudgery of much farm work. On the other hand, Goss and Rodefeld (1977) argued that mechanization has also resulted in increased farm size and labor displacement. Similarly, a USDA report recently asserted that "the major technological change which caused increased farm size was technological change biased toward machinery" (Lu, 1979:122).

Farm Workers

Farm workers have been marginal to American society at the same time as they have been central to the production process. Until the mid-1960s farm workers remained totally unorganized, unprotected by labor relations laws, and, generally, with few options for alternative employment. In 1940 the Interbureau Committee on Technology of USDA noted: "It has been pointed out that many tenants and laborers have been displaced from agriculture by the adoption of technological improvements. Certainly it is unlikely that their level of living has been improved" (1940:69). During the following decade, "the modernization of cotton production in the South, in large part a result of work done by land-grant colleges, displaced millions of people" (Paarlberg, 1981:132). Indeed, Goss and Rodefeld argued that agricultural mechanization has reduced the need for both hired and family workers and has, generally, had a greater impact on nonwhite than on white households (1977). Similarly, Boysie E. Day argued in a recent publication of the American Society of Agronomy that the agronomist "has brought about conversion of a rural agricultural society to an urban one. Each advance in his technology has sent a wave of displaced farm workers to seek a new life in the city and a flood of change throughout society" (1978:19). At the same time, agricultural mechanization has, as agricultural engineer W. E. Splinter (1980) argued, removed toil from farm work and increased the ratio of full-time to part-time farm and agricultural employees.

Only two fully detailed studies of the impact of agricultural technology upon farm workers exist, both of them the product of recent research by William Friedland and associates. The first of these studies was a retrospective study of tomato harvest mechanization in California (Friedland and Barton, 1975). In that study, the authors noted the shift from a predominantly male to a female labor force as mechanization progressed. In addition, they noted that from the point of view of labor, what these harvesting machines did was reproduce the labor relations of the assembly line in the field. Consequently, despite the sophisticated mechanical technology, "for most growers, workers are simply pairs of hands employed to perform a job" (Friedland and Barton, 1975:64).

In their study of lettuce harvest mechanization, Friedland, Barton, and Thomas (1978) argued that substantial labor displacement will occur as a result of lettuce harvest mechanization. Depending upon the number of farmers who adopt mechanization and the number of workers used on the machines, they projected that anywhere between 37.6% and 87.1% of the labor force will be displaced as a result of mechanization. Moreover, they noted that though jobs will be created in the construction of harvesting equipment, it is unlikely that these jobs will be open to the unskilled workers displaced by mechanization. Some lettuce harvesters may be employed in repair and maintenance of harvesting equipment; here too, however, the work requires high level skills, and it is unlikely that many lettuce harvesters will be able to make this transition.

Friedland, Barton, and Thomas also saw major changes in the characteristics of the labor force as a result of lettuce harvest mechanization. Specifically, they saw a shift toward (1) a substitution of female for male labor, (2) machine-paced operations, and (3) full-time employment of fewer persons, as "normal" employees of lettuce firms rather than as short-term seasonal workers.

Similar concerns were raised by Runyan, Bouma, and Peck (1975) in their discussion of increasing productivity in food wholesaling: "The adoption of highly advanced equipment raises both social and economic questions. The social questions are: (1) How will the people who are replaced by equipment earn a living; and (2) How long will it take for employees performing the routine tasks in highly mechanized or automated warehouses to rebel as they have in the auto industry?" (1975:11). These authors were so concerned with labor displacement and changes in skills that they proposed better supervision as a long-term substitute for automation.

Agribusiness

In contrast to the literature on farm workers, that on agribusiness has pointed to the benefits agribusiness receives from scientific research. As one USDA report recently noted, this has been particularly the case for animal agriculture (Penn, 1979). There, research findings and development have made possible integration, specialization, and concentration. As Sawyer observed in a discussion of the poultry industry, before integration "the farm chicken business was a sideline" (1971:35). Only after chicken production lines were industrialized and antibiotics were developed to prevent the spread of disease was it possible to engage in chicken production as a full-time business. Indeed, chicken raising has become so concentrated that virtually none of the approximately 300 trade journals published some years ago are still publishing today (Sawyer, 1971). The need and audience for such publications no longer exist. One irony of this rapid concentration in poultry production has been the sharp decline in poultry science departments at land-grant universities. The departments most responsible for the industrialization of chicken raising contributed to their own demise. A similar, though not as marked, trend toward concentration can be observed in the development of huge cattle feedlots run by large corporations.

Agribusiness has also benefited from the capital-intensive character of many of the products of agricultural research. In particular, input manufacturers (e.g., farm machinery, chemicals) and the processors and retailers of farm products have captured an increasingly large share of the agricultural dollar. It is now estimated that 40% of the value of agricultural products is added by input manufacturers, and 50% is added by processing, transport, and marketing. Only 10% of the total value is added on the farm (Lewontin, 1982). In addition, the percentage of the nonfarm share of retail food costs has risen from 51% in 1951 to 63% in 1980 (USDA, ERS, 1963; USDA, 1981).

Agribusiness has also indirectly benefited in that it can now demand more standardized products from farmers. Such uniform commodities may significantly reduce processing costs. At the same time, they may be no easier to cultivate and may increase waste at the farm level. In addition, the very practice of developing innovations that reduce production risks has made certain aspects of agricultural production more enticing to agribusiness (Perelman, 1982).

Rural Residents

As the products of agricultural science and technology have been adopted by farmers, they have certainly contributed to changing the face of rural America. The reduction in labor involved in farming has reduced the demand for some rural services. In particular, the number of agriculturally dependent places (e.g., counties, towns, and villages) has declined (Vogeler, 1981). Goldschmidt (1978) and others have argued that within the remaining communities, class differences have been exaggerated and the quality of life has declined. Goss and Rodefeld (1977) pointed out that rural to urban migration has generally been seen by agricultural economists as an adjustment process. If such is the case, they asserted rural areas should have less underemployment and higher income after outmigration. However, no evidence exists that would support the adjustment thesis.

On the other hand, certain segments of rural communities undoubtedly have benefited from the products of agricultural research. The remaining dealers for input supply manufacturers, as well as certain types of marketing facilities, have prospered and grown. In some cases, as the geographic center of production of a particular commodity has shifted, one rural area has prospered at the expense of another. In other cases, however, agricultural innovation has led to disastrous consequences for nearly everyone involved. Colman and Stockdale (1976) documented the failure of attempts to develop a sugar beet industry in upstate New York.

General Public

Arndt and Ruttan, among others, have pointed out that "in recent years there has been a proliferation of studies which indicate that returns to a great deal of investment in agricultural research has been two to three times higher than returns to other agricultural investment" (1977:4). In short, these studies have argued that money invested in public agricultural research brings returns to the public as a whole that are far larger than the initial investment. Yet, as Arndt and Ruttan themselves pointed out, few studies have taken into account the complex distributional effects wrought by the products of agricultural research. In fact, there has been a tendency to count only the growth benefits and to discount health and distributive justice (Kaldor, 1971).

Although, as a proportion of their total income, Americans pay less today for food than they did even 20 years ago, the nature of the food itself has been radically altered. The poor quality of tomatoes, grown for ease of shipment from California to New York, is legendary. Chemical additives, in a bewildering variety, are commonplace in the food eaten by most Americans. Sugar and salt, common ingredients in newly developed "convenience" foods, have come to play a major role and, clearly, an undesirable one in our diet. Much of the increased variety of food available is merely product differentiation rather than real choice.

Research has also altered the marketplace for both consumer and producer. Capital-intensive technologies have reduced the competition among buyers of farm products and among retailers of food products. The process was succinctly described by Alex McCalla: "As agriculture and agribusiness become more concentrated, the purely competitive assumption that cost reducing (efficiency) research is ultimately passed on to the consumer in the form of lower costs becomes more tenuous" (1973:1001). Indeed, an increasing share of the consumer's food dollar goes to processors and retailers.

In short, while the public may pay proportionally less today for food, it has done so only by accepting a number of problematic trade-offs. The case for agricultural research as a benefit to the public as a whole is by no means as clear as some would have us believe.

Foreign Groups

Most agricultural researchers see their research as likely to benefit foreign groups, institutions, or governments. However, here, too, the record is mixed. As we noted in Chapter 1, agricultural research in developing countries was initially oriented toward the production of export crops for metropolitan consumption. Even after independence, the bias toward export crops still remained. In the post–World War II period, Americans attempted to export to those countries not merely the products of agricultural research but the very institutional structures developed in the United States that appeared to be so successful in increasing agricultural productivity (Busch and Sachs, 1981).

One effect of this programmatic effort was the development of high yielding varieties of wheat and rice, the so called "green revolution." The green revolution emphasized the development of research institutions and technical research products as strategies for agricultural development, while it ignored the socioeconomic context. Andrew Pearse, in a volume summarizing a series of studies commissioned for the United Nations, has argued that the artifical separation of political from technical issues made it possible to avoid the issues of inequality and institutional change. In many cases, maldistribution of income has tended to be accentuated by the new technologies. As he stated, "the character of a technology and the strategy used to promote its adoption operate very forcefully upon the whole social system, including those factors upon which its distributive function depends" (Pearse, 1980:217). He concluded by arguing that only a peasant-based strategy will lead to both increased productivity and reduced inequality.

Conclusions

The perceptions of most practicing scientists and the informed empirical inquiries into who benefits from agricultural research present us with seemingly contradictory results. On the one hand, scientists appear to see many groups as benefiting equally from their research. On the other hand, studies of the actual consequences of agricultural research present a far more clouded picture. In the case of each beneficiary group, far more questions are raised than can be answered. One point clearly emerges: Nothing about any particular scientific inquiry insures that it will be of benefit to all, or even that the overall net social consequences will be desirable. We also are compelled to agree with agricultural economist Earl Heady, who warned, "funds for agricultural research and education are more likely to be guaranteed if allocation decisions [reflect] the fact that the welfare of a considerable [stratum] of the farm community is lessened by rapid farm technological change and if a larger amount of intellectual resources [are] diverted to solve or lessen the problems so created" (Heady, 1971:134).

9
Research for What?

From its inception, agricultural research has been goal oriented. Rarely have agricultural scientists engaged in research that merely satisfied personal curiosity. Indeed, many statements of goals have been written by nonscientists, and the most important of them have been incorporated into the various laws relating to public agricultural research.

The Hatch Act of 1887 was quite clear about the goals of research, specifying that agricultural research was "to aid in inquiring and diffusing among the people of the United States useful and practical information on subjects connected with agriculture, and to promote scientific investigation and experiment respecting the principles and applications of agricultural science" (Knoblauch, Law, and Meyer, 1962:219).

The Adams Act of 1906 further encouraged the conduct of original research directly related to agriculture. The Purnell Act of 1925 included yet another goal: "the development and improvement of the rural home and rural life" (Knoblauch, Law, and Meyer, 1962:222).

The Bankhead-Jones Act of 1935 supported "research relating to the improvement of the quality of, and the development of new and improved methods of production of, distribution of, and new and extended uses of markets for, agricultural commodities and byproducts thereof; and research relating to the conservation, development, and use of land and water resources for agricultural purposes" (Knoblauch, Law, and Meyer, 1962:223). Finally, the Agricultural Marketing Act of 1946 provided an even broader statement of the goals of agricultural research: "It is hereby declared to be the policy of the Congress to promote the efficient production and utilization of products of the soil as essential to the health and welfare of our people and to promote a sound and prosperous agricultural and rural life as indispensable to the maintenance of maximum employment and national prosperity" (Knoblauch, Law, and Meyer, 1962:225). This law listed a number of problems in need of research, including marketing, distribution, processing, human nutrition, the introduction of new plants, and the design and development of farm machinery. It specifically included "such investigations as have for their purpose the development and improvement of the rural home and rural life, and the maximum contribution by agriculture to the welfare of the consumer" (Knoblauch, Law, and Meyer, 1962:226).

Currently, the public agricultural research system employs a list of ten goals of agricultural research (USDA, Current Research Information System, 1978). The current goals classification system was initially developed as part of an attempt by the National Association of State Universities and Land-Grant Colleges (NASULGC) and USDA to improve the research priority-setting process (NASULGC and USDA, 1966). Over the years, the system has been modified slightly and a number of reports on priorities have been prepared (USDA, 1978; USDA, Science and Education Administration [SEA], 1980). In fact, elaborate priority-setting systems have been developed and used by research administrators. (For a description and critique of the priority-setting process see Ruttan [1982a].) Some systems employ numerical weighting schemes in an effort to objectify goal setting; others rely more directly on informed judgment. The current goals classification is also used to classify scientist-years and funds expended annually in agricultural research (USDA, SEA, 1981). Unfortunately, the goals are constructd so as to make interpretation of the data exceedingly difficult. (For more information see the Methodological Appendix.)

Occasionally, nongovernmental organizations also have expounded upon research goals. Goals similar to those of the federal government were incorporated into the report of the Commission on Country Life (1911). More recently, the American Society of Agricultural Engineers adopted its own goals program, whose "overriding goal" was "to lead in developing and implementing engineering-technology for agriculture that will optimally serve human needs and aspirations both now and in the future." The first goal in the list was "to develop public understanding of the consequences, both positive and negative, of existing and proposed engineering-technology-for-agriculture on human health, social structure, the environment, economy, and natural resources" (*Agricultural Engineering*, 1978:48). Clearly, such goals are laudable and should be further encouraged.

Unfortunately, almost all statements of agricultural research goals have two severe limitations. First, such statements rarely distinguish the relative importance of each goal. Second, they offer little evidence about the degree to which various institutions, scientists, and specific projects contribute to a given goal. In our study we attempt to address these issues from the perspectives of the scientists.

Scientists' Perceptions of Research Goals

While the individual objectives to be accomplished by a specific agricultural research project are generally quite narrow, scientists, both as individuals and as members of organizations, are likely to see their research as contributing to one or more broader goals. Few pieces of research, on the other hand, would be likely to contribute equally to the full range of goals that might be considered appropriate for agricultural research.

Using an abbreviated and modified version of the list currently employed by USDA (USDA, CRIS, 1978), we constructed a list of eleven goals for research. We then asked scientists to rate each goal in terms of: (1) its intrinsic importance as a goal for agricultural research and (2) the degree to which their own research contributed to that goal. In each case, scientists were asked to rate each goal on a scale ranging from 1 (of no importance) to 7 (of highest importance). This permitted us to examine how scientists assess their own work, as well as the breadth (or narrowness) of their view of the goals of agricultural research.

Mean scores for each of the eleven goals are presented in Figure 9.1. Development of new knowledge or improved methodology is ranked as clearly the most important goal for agricultural research. The second most important goal perceived by scientists is to increase agricultural productivity. Also important are the protection of crops and livestock and the reduction of production costs. The promotion of community improvement, improvement of marketing efficiency, and expansion of export markets rank lowest as research goals.

There is a strong relationship between those goals to which scientists see their research as contributing and those goals perceived as intrinsically important. Indeed, the average correlation between importance and contribution is 0.65. For this reason, only scientists' perceived contributions are reported below. One interpretation of this correlation is that scientists tend to undervalue the importance of those activities to which they themselves contribute little. This may be another factor contributing to the fragmentation of the agricultural research community.

As we have shown previously, however, overall means tend to conceal a variety of differences among scientists due to background,

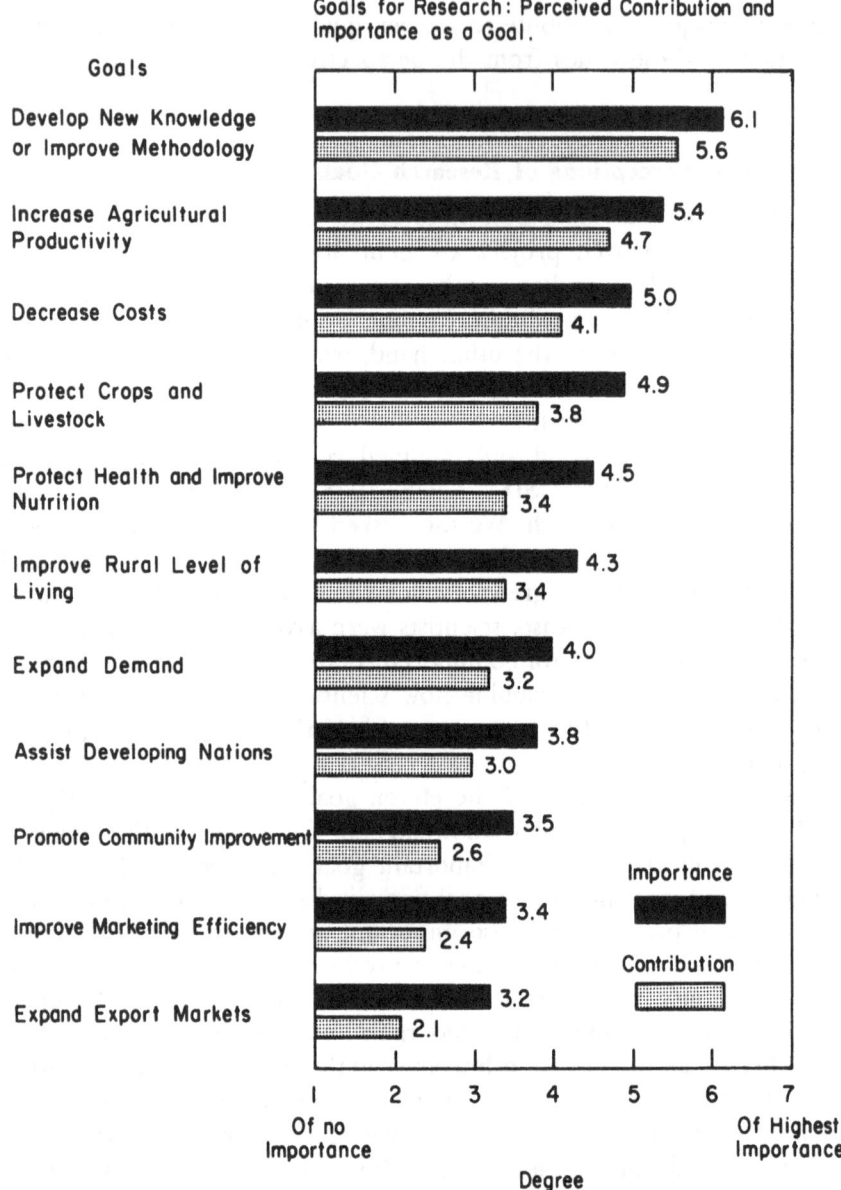

FIGURE 9.1
Goals for Research: Perceived Contribution and Importance as a Goal.

communication frequency, discipline, organization, and extraorganizational linkages.

Background

Those scientists from nonfarm backgrounds tend to see their research as contributing to somewhat different goals than do those

TABLE 9.1
Scientists' Contribution to Selected Research Goals by
Father's Occupation[a]

	Father's Occupation	
Goal	Farmer	Other
Increase agricultural productivity	5.1[b]	4.5
Decrease costs	4.6[b]	3.8
Improve marketing	2.6[b]	2.3
Aid developing nations	3.1[c]	2.9
Improve level of living	3.8[b]	3.2

[a]Mean scores based on seven point scale (1 = of no importance, 7 = of highest importance).

[b]Row pair differences, $p < .0001$.

[c]$p < .01$

scientists from farm backgrounds (Table 9.1). Specifically, scientists from farm backgrounds are likely to rate five out of eleven goals as more important than their nonfarm colleagues: increasing agricultural productivity, decreasing costs, improving marketing, aiding developing nations, and improving the level of living of rural communities. On the other hand, we found no association between a farm background and other research goals such as crop protection, health, and community improvement. This implies that, as more scientists from nonfarm backgrounds enter the system during the coming decades, there may be a marked shift away from certain traditional research goals.

Communication Frequency

Frequency of communication with various groups appears to have an equally important impact on scientists' contributions to various research goals (Table 9.2). Communication with scientists in one's department appears related to a perceived contribution to agricultural productivity, crop protection, decreased costs, and aid to developing nations. Communication with agricultural scientists outside one's department is also positively related to a perceived contribution to these goals. In addition, however, scientists who communicate more frequently with other agricultural scientists outside their departments are likely to be less concerned with the creation of new knowledge or improved methods.

TABLE 9.2
Contribution to Research Goals by Scientists' Communication Frequency with System Participants[a]

System Participant	Goals										
	Productivity	Protection	Decrease Costs	Demand	Marketing	Exports	Aid Developing Nations	Health	Improve Level of Living	Community Improvement	Knowledge
Scientists in department											
Rarely	4.2[b]	3.6[b]	3.7[b]	3.3	2.3	2.0	2.6[b]	3.4	3.1	2.6	5.5
Daily	5.1	4.2	4.4	3.3	2.4	2.2	3.1	3.6	3.5	2.6	5.7
Other agricultural scientists											
Rarely	3.9[b]	3.1[b]	3.4[b]	2.8[b]	2.1	1.8	2.6	3.3	3.2	2.5	5.4[b]
Daily	5.9	5.0	4.7	2.8	2.4	2.2	3.1	3.6	3.2	2.3	5.1
Other scientists											
Rarely	4.9[b]	3.8	4.4[b]	3.3	2.5	2.1	3.0	3.3	3.4	2.5	5.4[b]
Daily	4.1	4.4	3.8	3.1	2.9	2.4	4.1	4.0	3.0	3.1	5.9
Administration											
Rarely	4.4[b]	3.6[b]	3.8[b]	3.1	2.2[b]	1.9[b]	2.8[b]	3.4	3.1[b]	2.3[b]	5.6
Daily	5.4	4.4	4.9	4.0	2.4	2.6	3.4	3.5	3.1	2.1	5.7
Clients											
Rarely	4.2[b]	3.5[b]	3.6[b]	2.9[b]	2.0[b]	1.8[b]	2.8	3.3	2.9[b]	2.2[b]	5.6
Daily	5.9	5.4	5.3	3.4	2.3	2.7	3.3	4.0	4.6	2.9	5.6
Funding agencies											
Rarely	4.7	3.8	4.0	3.2	2.3	2.0	2.9	3.3	3.2[b]	2.4[b]	5.5
Daily	2.0	1.5	3.0	3.5	2.5	2.5	1.0	3.5	2.0	1.5	5.5
Extension staff											
Rarely	4.0[b]	3.4[b]	3.4[b]	3.0	2.1[b]	1.9	2.7[b]	3.3	2.9[b]	2.3[b]	5.0
Daily	5.6	4.9	5.1	3.3	2.7	2.2	3.1	4.0	4.2	2.9	5.4

[a] Mean scores based on a seven point scale (1 = of no importance, 7 = of highest importance).

[b] For column differences, p < .01.

Those scientists who communicate frequently with nonagricultural scientists appear to take a significantly different approach to their research. They are less concerned about agricultural productivity and decreasing costs and more concerned about the creation of new knowledge and improved methods. Communication with administrators reveals yet another pattern. Frequent communication with administrators is positively related to a concern with increased productivity, crop protection, decreased costs, improved marketing, increased exports, and aid to developing nations. In addition, there appears to be a curvilinear relationship between communication with administrators and concern for rural level of living and community improvement. In both cases, individuals with highest scores communicate occasionally with administrators. The import of this curvilinear relation is not clear.

Agricultural scientists who communicate frequently with clients see their research as contributing more to all goals *except* aid to developing nations, health, and the development of new knowledge. On the other hand, communication with funding agencies is significantly related only to a decreased concern for rural level of living and community improvement. Finally, communication with extension staff is positively related to all goals except increasing demand, increasing exports, health, and the development of new knowledge.

In summary, communication within the agricultural sciences and with clients tends to be associated with a contribution to the traditional goals of increased productivity, crop protection, and decreased costs. In contrast, communication with scientists outside the agricultural system is linked to a greater concern with the development of new knowledge.

Discipline

Significant disciplinary differences emerged in the degree to which scientists perceive their research to contribute to each of the eleven goals (Table 9.3). Horticulturists, plant pathologists, agronomists, and animal scientists are most likely to see their research as contributing to increased productivity, and social scientists and nutritionists are least likely. Not surprisingly, entomologists and plant pathologists see their research as contributing most to crop protection; social scientists and agricultural economists feel that they contribute least to this goal. In the case of reduced costs, horticulturists, animal scientists, and agricultural engineers top the list, with social scientists and food scientists at the bottom. On the other hand, food scientists see their research as contributing most to increased demand; social scientists appear to contribute the least to this goal.

The improvement of marketing efficiency appears to be the sole province of agricultural economists; no other discipline has a mean score above the midpoint of 3.5 on the 7 point scale. Similarly, few scientists in any discipline see their research as contributing significantly to the expansion of export markets. Assistance to developing nations fares only slightly better.

Research on the improvement of the level of living of rural America

TABLE 9.3
Scientists' Contribution to Research Goals by Discipline[a]

Discipline	Goals										
	Productivity	Protection	Decrease Costs	Demands	Marketing	Exports	Aid Developing Nations	Health	Improve Level of Living	Community Improvement	Knowledge
Agricultural Economics	3.6^{x}	1.6^{v}	3.9^{wx}	2.4^{ww}	4.4^{z}	2.6^{yz}	2.4^{wx}	2.2^{ww}	3.9^{y}	3.0^{y}	4.4^{y}
Agricultural Engineering	4.9^{yz}	3.8^{y}	4.8^{yz}	3.3^{wx}	2.5^{wx}	2.1^{xyz}	3.1^{xyz}	2.8^{wx}	3.7^{y}	2.9^{xv}	5.9^{z}
Agronomy	5.6^{z}	4.1^{y}	4.7^{xyz}	3.1^{wx}	1.9^{w}	2.0^{wxy}	3.3^{xyz}	3.2^{xv}	3.5^{xv}	2.5^{tuvwxy}	5.5^{z}
Animal Science	5.5^{yz}	3.7^{y}	5.1^{z}	3.4^{x}	2.5^{w}	2.0^{wxy}	2.8^{wxyz}	3.8^{y}	3.4^{xv}	2.2^{tuvw}	5.6^{z}
Basic Sciences	4.0^{x}	3.3^{xv}	3.1^{y}	3.1^{wx}	1.9^{w}	1.9^{wxy}	2.7^{wxy}	3.7^{y}	2.4^{w}	1.8^{t}	5.9^{z}
Entomology	5.1^{yz}	5.7^{z}	4.5^{xyz}	3.1^{wx}	1.9^{w}	2.0^{wxy}	3.2^{xyz}	3.8^{y}	3.1^{wxy}	2.3^{tuvwx}	5.5^{z}
Environmental Sciences	3.9^{x}	4.1^{y}	3.3^{ww}	2.5^{ww}	1.8^{w}	1.3^{w}	3.0^{wxyz}	3.0^{xv}	3.5^{xv}	3.1^{y}	5.8^{z}
Food Science	3.5^{x}	2.0^{ww}	3.1^{y}	5.4^{z}	3.3^{y}	2.8^{z}	2.8^{wxy}	5.3^{z}	2.8^{wx}	1.9^{tu}	5.9^{z}
Forestry	4.8^{y}	4.0^{y}	4.2^{xv}	3.7^{xv}	2.4^{w}	1.6^{wx}	2.2^{w}	1.6^{v}	3.2^{wxy}	2.7^{vwxy}	5.8^{z}
Horticulture	5.7^{z}	4.2^{y}	5.1^{z}	4.2^{y}	3.2^{xv}	2.5^{yz}	3.0^{wxyz}	3.6^{xv}	3.6^{xv}	2.8^{vwxy}	5.5^{z}
Nutrition	3.3^{x}	2.7^{wx}	2.9^{y}	3.0^{wx}	2.0^{w}	1.7^{wx}	3.4^{yz}	5.8^{z}	3.3^{xv}	2.9^{wxy}	5.9^{z}
Plant Pathology	5.4^{yz}	6.2^{z}	4.3^{xyz}	3.4^{xyz}	2.1^{w}	2.3^{xyz}	3.6^{z}	3.4^{xv}	3.1^{wxy}	2.1^{tuv}	5.7^{z}
Social Sciences	2.2^{w}	2.0^{ww}	1.8^{u}	1.7^{v}	2.0^{w}	1.6^{wx}	3.1^{xyz}	3.4^{xv}	5.7^{z}	5.7^{z}	5.6^{z}

[a]Mean scores based on a seven point scale (1 = of no importance, 7 = of highest importance). For all column differences, p < .001.

[t-z]Means within a column with different letters are significantly different, p < .01.

TABLE 9.4
Scientists' Contribution to Selected Research Goals by Region and Prestige of Institution[a]

| | Goal | | | |
	Increase Productivity	Decrease Costs	Increase Demand	Improve Level of Living
Region				
Northeast	4.4[b]	3.9[b]	3.1	3.4[b]
North Central	4.3	3.7	2.9	3.4
Southern	5.0	4.4	3.2	3.7
Western	4.9	4.1	3.2	3.0
Prestige				
High	3.9[b]	3.4[b]	2.9[b]	2.7[b]
Low	4.9	4.3	3.2	3.6

[a]Mean scores on a seven point scale (1 = of no importance, 7 = of highest importance).

[b]Column difference, p < .01.

and the promotion of community improvement appears to be overwhelmingly the province of social scientists. Finally, while all disciplines rate the development of new knowledge as their most important goal, it is significantly less important to agricultural economists than to those in other disciplines.

In short, disciplines vary widely in the degree to which they contribute to various research goals. Traditional goals such as increasing agricultural productivity are seen as the province of many disciplines, but more recently developed goals such as nutrition and community improvement tend to be relegated to one or two disciplines. Moreover, these disciplines tend to have fewer scientists and a more marginal role in the agricultural science system.

Organization

The context within which agricultural research is conducted appears to have little perceived effect upon the goals addressed by research. The only exceptions are slightly higher concerns for increasing demand and expanding export markets among government scientists as compared to their colleagues in universitites. On all other goals, however, no significant differences were observed. Among universities, however, there are both regional differences and differences between high and low prestige universities for several research goals (Table 9.4). Among scientists in southern experiment stations, greater emphasis appears to be given to research designed to increase productivity, to decrease costs, and to improve the level of living in rural America. Scientists in the western region are nearly as likely to work toward increased

TABLE 9.5
Correlations Between Selected Criteria for Promotion and Tenure and Contribution
to Selected Research Goals[a]

	Goals									
Criterion	Productivity	Protection	Decrease Costs	Demand	Marketing	Exports	Developing Nations	Health	Rural Level of Living	Community Improvement
Quality papers	.02	.07	.06	.07	.04	.06	.06	.10	.11	.05
In-house reports	.16	.05	.17	.04	.13	.07	.05	-.03	.21	.13
Government consulting	.05	.02	.06	.11	.09	.14	.07	.12	.09	.06
Private consulting	.07	.05	.09	.15	.11	.11	.08	.08	.14	.09
Teaching[b]	.11	.09	.05	.10	.04	.09	.12	.12	.11	.09
Extension[b]	.16	.09	.15	.13	.13	.05	.01	.03	.13	.05

[a]Pearsonian correlation coefficients (r). Values range from -1.0 (strong negative relationship) to 1.0 (strong positive relationship). A value of 0.0 is interpreted as the complete absence of any relation between the two variables. As the data is ordinal and the coefficients assume interval data, reported results are conservative. If $r > \pm .07$, then $p < .01$.

[b]Includes only university scientists.

productivity and decreased costs, but are least likely to see their research as contributing to an improved level of living for rural residents.

Institutional prestige also appears to play a role in scientists' perceived contributions to several research goals. Scientists in low prestige institutions are more likely to be concerned with increased productivity, decreased costs, increased demand, and improving the level of living in rural America. On all other goals, no significant difference was observed. These findings appear similar to those reported in Chapter 6, which suggested that scientists in the South are more concerned with immediate client needs and that scientists at high prestige institutions are less concerned with these kinds of needs.

The current policies for promotion and tenure appear to have a significant, though weak, effect upon most research goals (Table 9.5). Only two aspects of current promotion policy are unrelated to all the research goals: the publication of many papers and the receipt of a grant or contract. Similarly, only the goal of contributing to new knowledge appears unaffected by promotion policy. Scientists in institutions where high quality papers are important for promotion and tenure are more likely to do research contributing to health and nutrition as well as to improvement of the rural level of living. In those institutions where in-house reports are considered important, additional weight is given to increased productivity, decreased production costs, improved marketing efficiency, improved levels of living, and community improvement.

The weight given to consulting with both government and private organizations also appears to have an impact on the contribution scientists make to various research goals. Scientists in institutions in which consulting with government is considered important are more likely to be concerned with increasing demand, exports, and consumer health. Private consulting is positively related to a perceived contribution to demand, marketing, exports, and the level of living in rural communities.

Teaching evaluations are positively related to a perceived contribution to increasing productivity, demand, aid to developing nations, health, and the level of living of rural communities. Finally, extension evaluations are positively related to increasing productivity, decreasing costs, increasing demand, marketing, and improving levels of living.

Two conclusions may be drawn from a review of these relationships. First, they are relatively weak, which suggests that none of these evaluation measures are very effective in directing scientists toward particular goals. Second, the measure considered most important by scientists—the number of journal articles—is virtually unrelated to scientists' perceptions of their contributions to any goal. In contrast, extension evaluations, which are ranked relatively low in overall importance and are rarely used in USDA agencies, appear to have the greatest impact upon research goals.

Of course, the results reported here are of *perceptions* of scientists rather than an empirical study of various evaluation systems. An empirical study utilizing a quasi-experimental design might well yield different results. Nevertheless, these data suggest that *existing* evaluation systems are relatively ineffective in shaping research goals.

Grant Receipt

Extraorganizational factors also appear to influence scientists' relative weighting of many research goals. For example, recipients of National Science Foundation grants perceive themselves as contributing less to increasing productivity, decreasing costs, expanding demand, improving marketing, improving levels of living, or improving communities (Table 9.6). On the other hand, they see themselves as contributing significantly more to the development of new knowledge.

In the case of NIH grant recipients, a similar pattern is apparent. Recipients are more likely to see their research as contributing to

TABLE 9.6
Scientists' Contribution to Selected Research Goals by
Receipt of NSF or NIH Grant[a]

| | Organization | | | |
| | NSF Grant | | NIH Grant | |
Goal	Yes	No	Yes	No
Increase agricultural productivity	4.1[b]	4.8	3.5[b]	4.8
Decrease costs	3.2[b]	4.2	2.8[b]	4.2
Expand demand	2.8[c]	3.3	2.4[b]	3.3
Improve marketing	2.0[c]	2.4	1.6[b]	2.4
Expand exports	1.8	2.1	1.6[c]	2.1
Protect consumer health	3.4[b]	3.4	4.2[b]	3.4
Improve level of living	2.6[b]	3.5	2.6[b]	3.5
Improve communities	2.1[c]	2.6	2.1[c]	2.6
Develop new knowledge	6.0[b]	5.5	6.0[b]	5.6

[a]Mean scores on a seven point scale (1 = of no importance,
7 = of highest importance).

[b]Row pair differences, $p < .01$.

[c]Row pair differences, $p < .05$.

health and nutrition, as well as the development of new knowledge, but less likely to see their research as contributing to the other seven goals.

The receipt of grants and contracts from state and local governments also appears to influence scientists' perceptions of their contribution to several goals. Specifically, scientists with such grants or contracts are significantly more concerned with improving the level of living of rural America and promoting community improvement. The same is true for recipients of USDA grants and contracts, although to a lesser degree.

Recipients of commodity association grants are more likely than nonrecipients to see themselves as contributing to increasing productivity, protecting crops, decreasing costs, expanding demand, expanding exports, aiding developing nations, and improving consumer health and nutrition (Table 9.7). In contrast, private corporation grant recipients report greater contributions only to productivity, protection, costs, and demand. In the case of both commodity association and private corporation grants, recipients do not differ from nonrecipients in their concern for new knowledge, marketing, rural level of living, or community improvement.

TABLE 9.7
Scientists' Contribution to Selected Research Goals by Receipt of
Grant from Commodity Association or Private Corporation[a]

| | Organization | | | |
| | Commodity Association Grant | | Private Corporation Grant | |
Goal	Yes	No	Yes	No
Increase agricultural productivity	5.7[b]	4.6	5.5[b]	4.6
Protection	4.7[b]	3.7	4.5[b]	3.7
Decrease costs	5.2[b]	4.0	4.8[b]	4.0
Expand demand	4.0[b]	3.1	3.5[c]	3.2
Expand exports	2.5[b]	2.0	2.1	2.1
Aid developing nations	3.5[b]	2.9	3.0	3.0
Protect consumer health	3.9[b]	3.4	3.7	3.4

[a]Mean scores on a seven point scale (1 = of no importance, 7 = of highest importance).

[b]Row pair differences, $p < .01$.

[c]Row pair differences, $p < .05$.

Similar patterns emerge from other extraorganizational linkages (contracts, consulting, etc.), although the relationships are weaker. In general, scientists with relationships with NSF and NIH are more knowledge oriented, those with linkages to USDA and Interior are more community oriented, and those with connections with private groups are more production oriented.

In general, scientists perceive themselves as making the greatest contributions to the creation of new knowledge and the increase of agricultural productivity, and the least contribution to community improvement, marketing efficiency, and export markets. But scientists' contributions to research goals appear to be sharply divided along disciplinary lines. In addition, background, communication frequency, institutional location and prestige, and grant receipt all appear to play a role in scientists' choice and weighting of research goals.

Research Goals and Problem Choice

Research goals also influence the criteria one employs in choosing research problems. For example, one would expect that scientists who value knowledge most highly as a goal would also see the likelihood of making a contribution to scientific theory as a particularly important criterion in choosing a research problem. On the other hand, having chosen a problem because it fit certain criteria, scientists may come to modify the goals of their research. Thus, a scientist working on a relatively esoteric problem because it will advance scientific knowledge may properly exploit a practical application of that research, with a consequent change in his or her perceived contribution to, e.g., increasing agricultural productivity.

The relationship between relatively broad research goals and specific criteria for problem choice is a complex one that cannot be reduced to any simple causal relation. In order to highlight the key relationships, selected criteria for problem choice have been correlated with scientists' perceived contribution to each research goal and arranged in four groups (Table 9.8). Those items in group I include various "scientific" criteria for problem choice. Importantly, these are all negatively related or unrelated to "production" goals and, as expected, positively related to the quest for new knowledge. Of particular note are the extremely weak relationships between publication in scientific journals and any goal other than the creation of new knowledge. In contrast, those items in group II are strongly positively related to "production" goals but unrelated or negatively related to the creation of new knowledge. These findings once again illustrate the seeming antinomy between agricultural production and knowledge creation.

Each of the three items in group III is positively related to each of the research goals, although in each case the magnitude of the relationship differs. The creation of new methods, materials, and devices is most strongly related to increasing demand and productivity. A similar set of relations is observed for organizational priorities. In contrast, those scientists who were most concerned with improving the level of living of rural residents were also most likely to see importance to society as a significant criterion for problem choice.

The items in group IV appear to be related to research goals in a manner independent of each other or any of the criteria in the first three groups. The availability of research facilities is most strongly related to increasing productivity and reducing costs. It is virtually unrelated to marketing, exports, level of living, or community improvement, which reflects the minimal importance that physical facilities play in pursuit of these goals. Finally, scientists pursuing the less traditional research goals of improving the level of living of rural residents and improving communities were somewhat more likely to pursue hot topics than their colleagues. This may reflect the more tenuous nature of support for these goals.

Conflict and Consensus

The various official statements of research goals tend to assume that all may be equally pursued and that little or no conflict exists among them. Similarly, when respondents' scores on each of the various research goals are correlated, all the correlations are positive or not significantly different from zero. Scientists who assign an above average score to goal X are also likely to assign an above average score to goal Y and vice versa. This strongly suggests that scientists see no conflicts between goals.

Despite this, a number of observers have suggested the fundamental problem of goal conflict. Agricultural economist Earl O. Heady noted in 1949 that "the direction of technology advance might well differ depending upon the specific ends to be obtained" (1949:306), and that alternative goals were not without conflict. More recently, agricultural economist Harold O. Carter (1976) made a similar point. Wendell Berry (1977) argued that many of the current problems facing American agriculture are due to a fundamental conflict between the goals of health and productivity. Specifically, Berry argued that much current production-oriented research ignores questions of the health of people, or the land. As these goals—health and productivity— are pursued independently, and by members of different disciplines (e.g., agronomy versus nutrition), at the same time as some problems are "solved," others may be created.

The problem of goal conflict was brought home directly to us when we recently attended a multidisciplinary, international conference on sorghum, a grain used for both food and feed. One controversy revolved around breeding objectives. Breeders tended to emphasize maximum yield, whereas entomologists and pathologists were more concerned with insect and pathogen resistance, respectively. With different goals and little communication among these disciplines, certain "improved" varieties had the effect of creating new insect or pathogen problems that, in turn, had to be resolved by entomologists and pathologists.

Another major issue revolved around the food-versus-feed orientation of scientists. Many scientists interested in sorghum as feed were far less concerned with food quality or nutritional character than those scientists with a food orientation. Furthermore, it appeared that maximizing nutritional value might reduce both yield and the gastronomic quality of the various dishes prepared with sorghum. In

TABLE 9.8
Correlations Between Selected Criteria for Problem Choice and Research Goals[a]

Criterion	Goals										
	Productivity	Protection	Decrease Costs	Demand	Marketing	Exports	Aid Developing Nations	Health	Improve Level of Living	Community Improvement	Knowledge
Group I.											
Contribution to theory	-.06	.03	-.12	-.07	-.15	-.03	.04	.10	-.07	-.05	.30
Publish in professional journals	.02	.03	.01	.06	-.06	.00	.06	.12	-.03	-.07	.15
Scientific curiosity	-.11	.03	-.15	-.08	-.12	-.05	.05	.07	-.07	-.01	.22
Group II.											
Client needs assessment	.34	.18	.38	.20	.26	.19	.11	.09	.31	.18	-.05
Clientele demands	.26	.16	.32	.18	.26	.17	.06	.03	.21	.13	-.14
Feedback from extension	.29	.22	.33	.16	.20	.09	.11	.09	.23	.18	-.10
Marketability	.28	.13	.35	.31	.27	.24	.15	.08	.26	.15	-.03
Publish in farm journals	.29	.14	.36	.26	.28	.20	.08	.08	.24	.17	-.03
Publish in bulletins	.26	.08	.30	.14	.25	.12	.06	-.04	.29	.23	-.10
Group III.											
Importance to society	.23	.12	.23	.15	.12	.13	.23	.21	.30	.22	.11
Methods, materials, devices	.23	.18	.22	.27	.10	.15	.20	.12	.15	.08	.17
Organizational priorities	.22	.14	.21	.23	.20	.20	.12	.11	.12	.08	.02
Group IV.											
Availability of facilities	.20	.16	.20	.19	-.04	.06	.10	.14	.03	-.03	.12
"Hot" topics	-.03	-.07	.05	.10	.15	.11	.01	.05	.15	.16	.00

[a]Pearsonian correlation coefficients (r). Values range from -1.0 (strong negative relationship) to 1.0 (strong positive relationship). A value of 0.0 is interpreted as the complete absence of any relation between the two variables. As the data are ordinal and the coefficients assume interval data, reported results are conservative. If $r > \pm.09$, then $p < .001$.

short, the goals and objectives of the various disciplines could not be maximized simultaneously.

Even within disciplines, goal conflicts may arise. For example, in recent years a number of reports on nutrition research have been published by various government agencies (e.g., OTA, GAO). One of the more recent reports argued:

> The United States has gradually shifted its nutrition focus away from nutrient deficiency questions, and toward biochemical functions of the nutrients and undernutrition in developing countries. This shift has left a vacuum in domestic human nutrition research. . . . The Federal Government has failed to adjust the emphasis of its human nutrition research activities to deal with the changing health problems of the people of the United States (OTA, 1978:3–4).

Particular emphasis was placed on the lack of research on the role of nutrition in degenerative diseases, including heart diseases.

These findings were controversial enough to prompt a strong rebuttal by the chairman of a major nutrition department. Alfred E. Harper (1979), writing in *Nutrition Today*, listed what he saw as contradictions in the OTA report. He also argued that the OTA report assumed a relationship between nutrition and degenerative diseases, although no evidence existed to support that relationship.

Intradisciplinary goal conflict was also evident in a recent debate within agricultural economics, when ERS economists attempted to define research goals more clearly. In the report of that effort, one paper emphasized equity while another paper emphasized the more traditional goal of efficiency in the farm sector (USDA, ERS, 1977; see also West, 1973).

In summary, official statements and scientific pronouncements notwithstanding, in a world with limited resources and time, some research goals must necessarily conflict. This suggests that any attempt to maximize one goal may seriously reduce the likelihood of realizing others or optimizing the full set of goals.

Goals and Beneficiaries

The data reported here on perceived beneficiaries (Chapter 8) and on research goals point to a fundamental anomaly in scientists' understandings of the role of agricultural research in the larger society. On the one hand, scientists see most groups of beneficiaries as benefiting about equally from their research. On the other hand, scientists see certain goals as significantly more important than others in the conduct of their research. This would be relatively unproblematic if there were no link between the maximization of particular research goals and the flow of benefits of research to certain groups. However, this is certainly not the case. For example, successful efforts to increase agricultural productivity are likely to benefit: (1) those farmers who are early adopters of the technology, (2) processing and marketing firms that are able to purchase agricultural commodities at lower prices, and, perhaps, (3) consumers. On the other hand, such efforts are less likely to benefit those farmers who, for one reason or another, are late adopters. Finally, in the event that food quality and nutritional goals are neglected, or in oligopolistic product markets, little or no benefit may accrue to consumers.

Similarly, the promotion of community improvement may cost some groups and benefit others. Elimination of nonpoint source pollution from fertilizer runoff is likely to cost farmers but benefit other rural residents. Similarly, the expansion of exports of agricultural commodities, if accomplished through government subsidy, may benefit farmers at the same time as it raises prices for consumers. In fact, even the assistance provided to developing nations in agricultural research has its costs, which must be borne by the general public. Crop and livestock protection may be effectively accomplished through the use of chemical sprays, but such sprays may also increase health hazards to farm workers and the general public. On the other hand, protecting consumer health may well increase the economic risks to producers of crops and livestock. Without question, demands for health and environmental guarantees have resulted in increased costs associated with product testing and development for chemical companies.

In short, the pursuit of every goal brings with it both costs and benefits. On the surface, it might appear that the solution is to develop a system that maximizes the benefits and minimizes the costs. However, not even such a system would be entirely satisfactory, for it would answer only the question, "Research for what?" It would leave unanswered the question, "Research for whom?" From the pursuit of any given goal, certain groups and individuals are likely to derive benefits, whereas other groups and individuals are likely to incur costs. No simple economic cost/benefit analysis can resolve this fundamental problem.

A thorough examination of the complex trade-offs among goals for agricultural research is not possible in this volume. Nevertheless, establishing goals and priorities clearly does entail trade-offs and questions of who shall benefit from the agricultural research enterprise. Such questions highlight the need for a more informed agricultural research policy.

10
Science, Agriculture, and Emerging Issues for Agricultural Research

During the last decade, a number of issues have emerged regarding both science and agriculture, the two broader systems in which the agricultural sciences are embedded. These issues have wide ranging implications for the goals, processes, products, and beneficiaries of agricultural science. Both science and agriculture have changed dramatically in this century, and consequently, the basic assumptions of both have required a reexamination.

Science

Until recently, anyone asking, "What is science for?" would have been categorized as ignorant, uninformed, or a devil's advocate. As early as the seventeenth century, the philosopher-scientist Francis Bacon had explained that science was to be used for the merit and emolument of man. Scientists such as Kepler, Newton, and Pasteur demonstrated that science and technology were destined to be the chief architects of human progress. Fundamental to this view was the acceptance of science as an autonomous institution, creating knowledge that was an accurate representation of the world, and hence not subject to external manipulation. Societal influences impinged upon the world of science only to the extent that nonscientific criteria were employed in the allocation of research funds (Nicholson, 1977). Exceptions were viewed as merely evidence of paradigmatic immaturity (Nagi and Corwin, 1972). As a consequence, during the last century, huge sums of money were devoted to scientific research, development, and training throughout the world in the confident belief that science is an indisputable good that creates knowledge free from the constraints of human circumstance.

Recently, however, the assumptions, as well as the products, of science have come under attack. First, the claims of science to objective knowledge have been challenged. Several social scientists and philosophers (e.g., Busch and Lacy, 1981; Rosenberg, 1976; Whitley, 1974) have asserted that science shares the ideologies prevailing in a given society at a given time and that these ideologies affect a number of aspects of science. They have argued that values and ideologies, as well as economic factors, affect: (1) the organization of the sciences into broad areas and specific disciplines, (2) scientific research styles or orientations, (3) the degree of permissible deviation from the established scientific orthodoxy, (4) the way in which research results are reported, and (5) even the choice of research problems. One biologist (Luria, 1973) noted that researchers have become opportunistic research entrepreneurs in response to increases in the scale of their activities and the creation of direct state funding. This, he acknowledged, is of little consequence if it serves merely to direct scientists to research problems of social significance. However, "the danger is that a man may come to identify his scholarly function with the specific goals of certain sectors of the power structure" (Luria, 1973:80).

These challenges to an autonomous, objective science have been bolstered by two related changes: the decline of the belief in the unity of science, and the changing role of science in society. With the decline of the belief in the unity of science (Kuhn, 1970), perfect knowledge was no longer viewed by either scientists or those who studied them as a valid goal. It could now be argued that knowledge of certain aspects of the world must be developed at the expense of knowledge of others. Many separate sciences could develop out of many cultural systems. These sciences would not produce contradictory results, but would deal with different aspects of the natural world.

The role of science in society has also undergone dramatic change. The status of science has changed from that of a peripheral to that of a core institution, with a corresponding change in the role of science in the larger society. Initially, science was seen as a liberating force, freeing humanity from the bonds of tradition and scholasticism (Maritain, 1944; Rose and Rose, 1976), while the scientist was seen as an atomistic entrepreneur or a self-employed master craftsman, pursuing what he felt to be important, and confident that it would be a building block of knowledge.

With the development of the corporate form, science became corporatized; scientists became employees of the state and industry.

Moreover, this shift from Cartesian to Baconian science (Busch and Lacy, 1981; Haberer, 1969) changed the nature of scientific knowledge: "The dominant mode of production of scientific knowledge has become that of knowledge-as-commodity, as a marketable good with a cash value" (Rose and Rose, 1976:15). Copyright and patent laws are just two indices of this current state of science.

The increase in scale, the development of scientific "manpower," and the increasing centrality of the relations between science and the state have meant that science as an institution has become less autonomous. Moreover, both the state and the large corporations have become increasingly reliant upon scientific knowledge to bolster and to maintain their legitimacy. Indeed, some writers have argued that the claim of scientific autonomy has become an ideology designed to further certain class interests (Habermas, 1970).

In addition to changes in the assumptions of science and its role in society, science has been confronted with several practical considerations. First, in the middle 1960s, came questions about how many scientists we really needed and gloomy calculations about future reductions in the growth rate of investment in science. Then came a growing concern about the adverse consequences of our use of science and technology, from environmental pollution and genetic engineering, to the invasion of privacy by computers, nuclear radiation, and the awesome techniques of warfare. Finally, there was disappointment because science often promised more than it delivered. Several groups in society reached the conclusion that science was not the sole or supreme tool for improving the world. Even former president Carter, in a glowing endorsement of science presented to the National Academy of Science in 1979, cautioned that "the tree of science is always beautiful but its fruits can be bitter as well as sweet. Our task is to nourish the tree and harvest the fruits that are sweet" (Culliton, 1979:480).

U.S. Agriculture

Paralleling the changing assumptions and role of science, radical changes have taken place in U.S. agriculture, particularly since the end of World War II. These changes have been caused in part by an enormous growth in foreign demand for food, feed, and fiber, by the thrust of the research and extension programs of the agricultural research system, and by changes in U.S. agricultural and monetary policies.

A detailed examination of these underlying factors reveals that foreign demand accounted for 10% of all farm products marketed in the 1950s, but for 30% by the end of the 1970s. U.S. monetary policies in 1971, permitting the dollar to float in foreign exchange markets, contributed markedly to the increased demand by lowering U.S. product prices and making U.S. farm exports even more competitive (Bergland and Sechler, 1981).

At the same time, the rest of the world's self-sufficiency in grains and oilseeds (crops that account for three-fifths of total world agricultural production) dropped from 98% in 1950 to 90% in the late 1970s. This drop has contributed to a greater reliance on U.S. agriculture by foreign countries. In the early 1950s, the rest of the world depended on the U.S. for 2% of its agricultural supplies; in the late 1970s, for 11%. With projections of a widening gap between foreign production and demand, it is estimated that by 1985 the U.S. may be supplying 15% of the rest of the world's food, feed, and fiber.

During the fifty years before this large increase in foreign demand, the United States was viewed as having virtually limitless potential. There were years when the farm sector was operating at 75% of its capacity. During the 1960s, as many as 62 million acres, nearly one-fifth of U.S. cropland, were held out of production in an attempt to bring a balance between supply and demand. In the 1970s, U.S. agriculture entered what many people regarded as an era of limits and critical choices, requiring significant adjustments in the use of resources. The days of surplus resources and production were replaced by conditions under which production and demand were more closely in balance and resources were more fully utilized. Virtually all of the U.S. cropland previously idled through government programs was returned to production by the end of the 1970s.

While demand for U.S. exports was increasing and the availability of good land was decreasing, the nature of the farm sector was also experiencing major changes. Farmers continued to adopt technological advances and new practices based on intensive use of energy and petroleum-based chemicals, to develop land, and to increase land holdings. As a consequence, farms today are fewer, larger, more specialized, and more capital intensive. They are more dependent on industrial inputs and many are even more highly in debt. Under the new definition of a farm as a place that has agricultural product sales of $1,000 (Bergland and Sechler, 1981), the number of farms fell to 2.37 million in 1979, down from nearly 7 million in the 1930s. Lewontin (1982) notes that much of the decline was due to the decline of farm tenancies. The farm population has also dropped dramatically from 32 million in 1920 (30% of the total population) to 6.2 million in 1979 (3% of the population) (U.S. Department of Commerce, Bureau of the Census, 1979). Within this declining group of farms and farmers, a relatively small number of large farms are the major producers of food and fiber. In 1978, just 7% of the farms generated 56% of the total value of all food and fiber production and controlled well over half of the farmland acreage in the country.

This expansion of farm size has been an important aspect of past gains in production efficiency. Economies of size, stimulated in part by public research on such scale-biased technology as large equipment and capital-intensive tools, have permitted farmers to reduce unit production costs by increasing farm size. Other research has removed some of the biological, technical, and managerial constraints on large-scale specialized production (e.g., development of improved poultry disease-control techniques that enabled large-scale production of broilers). Several studies have reaffirmed that the long run average per unit costs for farms decrease rapidly as farm size increases from relatively small sizes. However, since many primary farms (gross sales of at least $40,000) have reached or surpassed the size needed to attain most economies of scale, further expansion may yield little benefit in improved efficiency (Bergland and Sechler, 1981; Faux, 1973; Madden and Parterheimer, 1972; Miller, Rodewald and McElroy, 1981). In fact, it has recently been argued that continued concentration of production capacity may inhibit the flexibility required to respond to changing conditions and needs. In addition, these large farms have relied heavily on inexpensive, energy intensive technologies and petroluem-based chemicals that have risen dramatically in cost and have experienced disruptions in supply during the 1970s. Although continued use of these technologies was economically unattractive during this period, large farmers, already heavily committed, continued to invest in these high-cost production inputs. Consequently, they became heavy users of borrowed capital, which also increased in cost during the 1970s and early 1980s.

Annual cash obligations are now a high proportion of gross receipts on all farms, and the proportion grows as the size of farms increases. This has resulted in increasing vulnerability to unstable commodity prices, a common occurrence during the past decade. Indeed, during the 1970s, prices fluctuated in five years from a postwar low to an all time high. The interlocking complexities of today's agricultural environment have meant that unanticipated shifts in a single element, such as the weather or the cost of a production input like petroleum, can have widespread ramifications for prices, supplies, and market activities worldwide.

Related to the changing U.S. production capacity and efficiency is the growing recognition of the constraints on nonrenewable natural resources as production inputs and of the necessity to maintain minimum levels of environmental quality. As the capacity of the U.S. production system is pushed to its limits, the physical limits of our natural resource base are becoming more fully appreciated. First, there are the growing and competing demands for other uses of prime farm land, water, forests, and other natural resources. Second, as remaining reserves of readily available, relatively fertile land are depleted, the expansion of agriculture will mean the elimination of previously fallow acreage and movement onto less productive and more fragile soils. This will likely entail greater production inputs, such as irrigation water and fertilizers, and increased risks of erosion (more than one-half of all erosion occurs on cropland) and other environmental damage. Such efforts imply a potential loss in future productive capacity, and increase the vulnerability of agricultural production to minor climatic variations (Bergland and Sechler, 1981; Lee, 1978; National Association of Conservation Districts, 1980).

Dwindling water supplies in some regions, water pollution, potential environmental and health hazards related to pesticides and herbicides, and other health hazards from past and current production practices have become increasingly problematic. Consequently, conservation of resources and protection of the environment—particularly land, water, and, more recently, energy—have received increasing attention. Enlightened definitions of efficiency have encompassed the day-to-day aspects of conservation, but serious debates regarding long- and short-term conservation are being raised and have implications for longer term food security. Several writers, such as Bergland and Sechler (1981), Friedland, Barton, and Thomas (1978), Berry (1977), and Hightower (1973), have argued that current agricultural practices have contributed to environmental degradation and may be inappropriate for ecological sustainability.

The growth in the specialized, capital intensive, concentrated farm sector is matched by a growing concentration throughout the entire food industry. Agricultural input suppliers, the agricultural marketing system, food processing, distributing, and retailing are all increasingly concentrated. With sizeable cost increases for such items as labor and energy, the processing and distribution sectors now account for a substantially larger portion of the consumer food dollar than the farm sector. In addition, agricultural production is increasingly seen as the source of raw materials for a food system dominated by processing, distribution, and marketing sectors, which increasingly make basic decisions concerning the kinds and volume of food products to be produced. Consequently, food and fiber prices are more likely to depend on improved productivity and efficiency in the nonfarm sector.

Finally, the agricultural sector has been confronted with increasing consumer interests and concerns about the products of this system. The economic concentration in marketing, processing, and distribution industries has raised concerns about the possible adverse effects of monopolistic power on both consumers and producers. Furthermore, changing farm technology, new hybrids bred for crop storage and shipment, and widespread use of pesticides, as well as increased processing of raw foods and numerous food additives, have led to concerns about the nutritional quality and safety of food. Consumers have begun to challenge agricultural research agendas that ignore nutrition and health and have demanded increased government regulation to provide for public health and safety. Finally, the public has begun to raise broad questions about the fundamental goals for the nation's food and agricultural system, including such issues as equity, efficiency, resilience, flexibility, conservation, and consistency with other objectives of U.S. society.

Agricultural Scientists' Response

These enormous changes in both the scientific and the agricultural communities suggest that major shifts may be required in publicly supported agricultural research. Farm production technologies that are more energy and capital efficient, as well as less energy and chemical intensive, are apparently needed. There is a need to explore the long-run potential of alternative technologies and unconventional production practices as ways to reduce production costs and dependence on scarce nonrenewable resources. It has also been suggested that new technologies and approaches be developed to reduce costs and increase the efficiency and economic viability of smaller and medium-sized farms, thereby promoting a more diverse, potentially more resilient and competitive agricultural system (Bergland and Sechler, 1981). Finally, research should assess the environmental, health, and nutritional hazards and benefits from food production and processing practices.

Agricultural scientists in many disciplines have begun to raise these same issues. Several researchers in our survey volunteered their observations about research issues. One scientist commented that "the academic community has not anticipated issues until they reach the crisis stage." This was echoed in a biochemist's remarks that

changes in production methods will be necessary owing to dwindling oil supplies. *Now* is the time to be concerned about twenty-five years in the future. Alternate methods of fertilization should be sought, alternate methods of pest control should be sought (genetic, microbial, etc.) and conservation of agricultural resources should be paramount.

Another agricultural scientist noted:

Agricultural research puts too much emphasis on applied research and neglects theoretical basics. It has been too subservient to pressure groups in Congress, business and major farm groups. It treats people-needs as minor; fails to fund home economics research adequately. Cows and pigs count more than people. Tomatoes look better than they taste. . . . Agricultural research has largely neglected climate research, plant diversity and many other critical issues too.

Another biochemist suggested the need to address

whether we have the capacity to produce agricultural commodities to satisfy our requirements for food, feed, fiber and export and to be used as renewable fuel to replace petroleum for industrial product manufacture. Research needs to determine whether commercial crops can fulfill both needs or whether unconventional crops need to be developed to fulfill certain product requirements. Research for improving nutritional quality and safety of protein for direct human consumption also requires a primary effort.

An agricultural economist observed that the "agricultural sciences are changing from farm orientation to resource management and planning emphases. Seeing 'agriculture' as more than farming is essential to continued evolution of work in the whole land management field." Finally, in conjunction with the need for this broader planning orientation, a researcher noted that "emphasis on policy concerns of stability, equity and distribution should be important factors in selecting research areas in agricultural policy and international trade."

To examine the extent to which agricultural scientists are currently addressing a number of these issues, we asked scientists to indicate the degree to which their research reflected five broad groups of issues of increasing current interest among scientists, administrators, and the public. These issues had emerged from a series of in-depth interviews with agricultural scientists at several institutions, and in part reflected changes in the nature and variety of organized interests concerned with agricultural research (Busch and Lacy, 1981). They included environmental issues (e.g., soil erosion, pest management); world food crisis (e.g., problems of food sufficiency and distribution); energy issues (e.g., fossil fuel usage, energy efficiency); consumer issues (e.g., nutrition, consumer health); and alternative approaches to agriculture (e.g., organic farming).

Of these five groups, scientists most frequently perceive environmental issues as reflected in their research. Their mean score is 4.4 for environmental concerns on a scale ranging from 1 (rarely) to 7 (frequently). These are the only issues that received a mean score above the midpoint on the scale. Scientists view their research as reflecting the world food crisis ($\bar{x}$ = 3.6) and energy issues ($\bar{x}$ = 3.5) to a lesser extent. Consumer issues are fourth among the groups ($\bar{x}$ = 3.2). Finally, scientists see their research as reflecting alternative approaches to agriculture only to very limited degree ($\bar{x}$ = 2.6). In fact, a scientist at one university wrote: "I don't believe that anyone interested in small or organic farming could get tenure here as he or she would be considered way off the beaten path. One possible help would be federal grants tied strictly to small farm, 'appropriate technology,' but administered through rural sociologists or interdepartmental groups of this orientation."

These overall evaluations, however, tend to conceal a variety of differences among scientists that affect their perceptions regarding these research issues. In previous chapters we have seen that family background, research orientation, communication frequency, academic discipline, research organizational membership, and contact with extraorganizational associations and institutions influence criteria for research problem choice, as well as goals and perceptions about beneficiaries. It is equally important to examine how these same factors may be related to research involvement with these five agricultural issues.

Background and Research Orientation

Scientists whose fathers were farmers are more likely than other agricultural scientists to indicate that their research reflects food issues ($\bar{x}$ = 3.9) and energy issues ($\bar{x}$ = 3.8). These are precisely the issues that have the most immediate and direct implications for farmers. In contrast, there are no differences between these two groups on the other issues.

Age is also relevant; older scientists are more likely to report that world food issues are reflected in their research. On the other hand, younger scientists are more likely to see their research as responding to recently emerging energy and environmental concerns. This is consistent with the younger scientists' tendency to choose "hot topics" and their hypothesized speedier acceptance of new ideas (Chapter 3).

Earlier, we demonstrated that scientists' orientation to basic or applied research or development was significantly related to a variety of criteria used in establishing research agendas (Chapter 3). To a lesser extent, this research orientation is also related to the degree to which scientists' research reflects the five issue groups. There is a significant negative correlation between the percentage of a scientist's research that is basic and the degree to which that scientist's research reflects energy issues ($r = -0.19$, $p < 0.001$), environmental issues ($r = -0.11$, $p < 0.001$), and alternative approaches to agriculture ($r = -0.09$, $p < 0.001$). In contrast, there is a positive correlation between the percentage of applied research conducted by scientists and their perceived involvement in these same three issues. This contrast raises some potentially serious questions about the probability of adequately addressing these issues. The most promising approaches are likely to require substantial, long-term, basic science inputs; yet, basic scientists in agriculture are least likely to view their research as addressing these issues.

Communication Frequency

As we have seen in previous chapters, informal communication among scientists is an extremely important dimension of the research process. Frequency of communication about research with various participants in the research system is positively related to the degree to which scientists perceive that their research reflects four of the five issues. Only consumer issues are unrelated to communication frequency with any group. However, communication frequency with agricultural scientists outside one's department, administrators, clients, and extension staff is positively related to energy, environmental, and food-related research and alternative approaches to agriculture (Table 10.1). For example, scientists with daily interaction with these four groups on research issues rate the degree to which their research reflects environmental issues at least 1.5 points higher on the average than those interacting rarely (7 point scale). Although the nature of the communication and the meaning of the research involvement in these issues needs to be further assessed, this relationship suggests the importance of promoting interpersonal contacts for research in these areas.

Academic Discipline

Perhaps the most interesting relationships are those between these five issues and academic disciplines. As we noted earlier in this chapter, the environmental issues are perceived as most frequently reflected in agricultural scientists' research ($\bar{x} = 4.4$). Not surprisingly, environmental scientists score highest on this item ($\bar{x} = 6.0$). Of equal importance is the high rating of environmental issues by entomologists,

TABLE 10.1
Correlations Between Degree to Which Research Reflects Current Issues and Communication Frequency with Selected System Participants[a]

Current Issues	Participants				
	Agricultural Scientists Outside Department	Non-Agricultural Scientists	Administrators	Clients	Extension Staff
Environmental issues	.20	.08	.14	.14	.14
World food issues	.20	.00	.08	.03	.10
Energy issues	.17	.02	.16	.19	.17
Consumer issues	.03	.01	.04	.04	.05
Alternative approaches to agriculture	.16	.04	.09	.06	.13

[a]Pearsonian correlation coefficients (r). Values range from -1.0 (strong negative relationship) to 1.0 (strong positive relationship). A value of 0.0 is interpreted as the complete absence of any relation between the two variables. As the data is ordinal and the coefficients assume interval data, reported results are conservative. If $r > .08$, then $p < .001$.

FIGURE 10.1
Degree to which Research Reflects Environmental Issues
by Discipline.

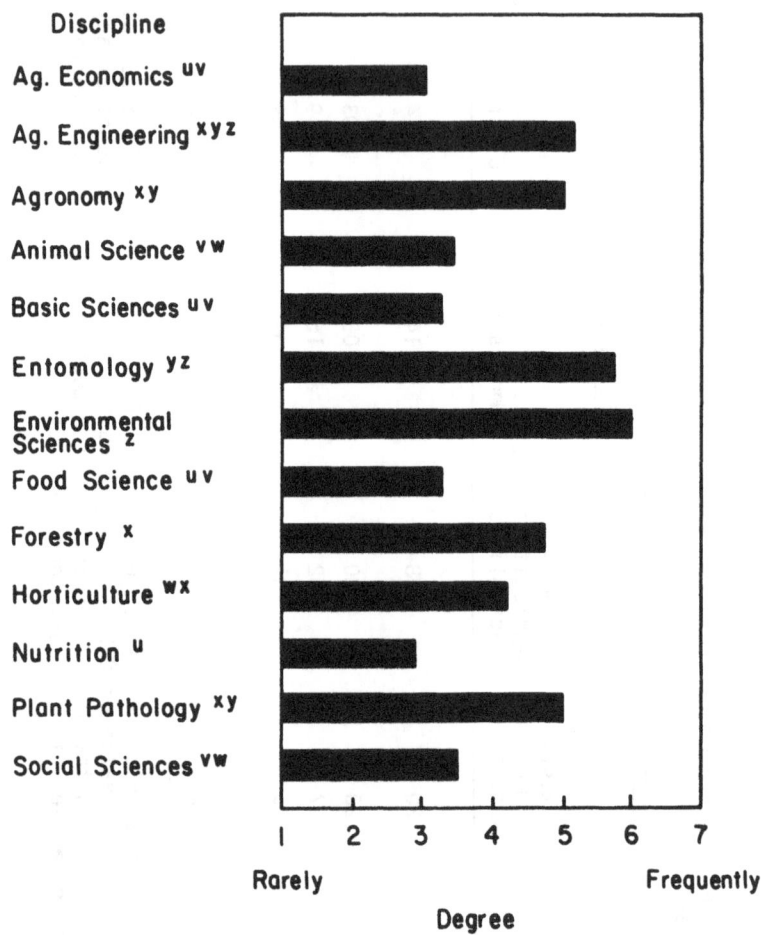

Discipline

Ag. Economics ᵘᵛ

Ag. Engineering ˣʸᶻ

Agronomy ˣʸ

Animal Science ᵛʷ

Basic Sciences ᵘᵛ

Entomology ʸᶻ

Environmental Sciences ᶻ

Food Science ᵘᵛ

Forestry ˣ

Horticulture ʷˣ

Nutrition ᵘ

Plant Pathology ˣʸ

Social Sciences ᵛʷ

| 1   2   3   4   5   6   7 |

Rarely   Frequently

Degree

ᵘ⁻ᶻ Disciplines with different letters are significantly different,
p < .01.

paralleling the substantial shift toward integrated pest management
over the last several years. Nutritionists rate this item lowest ($\bar{x} =$
2.9), implying that little work in that discipline has focused upon
environmental issues. Food scientists also score this item quite low,
apparently rejecting the connection between potential presence of
toxic substances in foods and the environment (Figure 10.1).

FIGURE 10.2
Degree to which Research Reflects World Food Crisis
by Discipline.

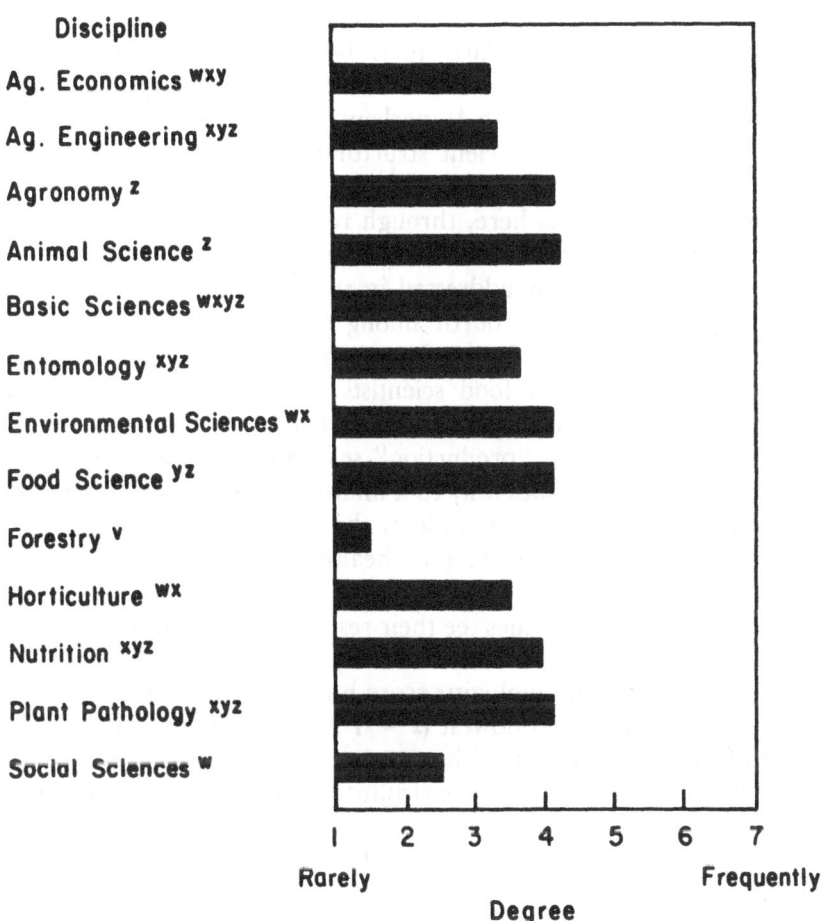

ᵛ⁻ᶻ Disciplines with different letters are significantly different,
p < .0I.

There is considerably less disciplinary variation in scientists' perceptions of the degree to which their research reflects the world food crisis (Figure 10.2). Animal scientists see these issues as more important than do those in other fields, although the mean scores for several other disciplines are not significantly lower. Of some concern here is the slightly lower score on this item among the basic scientists and the very low score of those in the social sciences. Given the mounting evidence for the importance of marketing, processing, social, and cultural issues in the world food crisis, the relatively low social science research effort suggests a need for policy change.

The mean score for all scientists on the degree to which research reflects energy issues (Figure 10.3) is 3.5. However, agricultural engineers clearly see their research as reflecting energy issues to a much greater degree ($\bar{x} = 5.4$), perhaps in response to the perceived need for more energy-efficient structures and farm machinery. On the other hand, the low mean score for the basic sciences ($\bar{x} = 2.6$) is problematic, for it is here, through research on nitrogen fixation, photosynthesis, and other fundamental biological processes, that energy issues may be effectively addressed in the long run.

Consumer issues rank fourth among the groups in the degree to which they are perceived as reflected in research ($\bar{x} = 3.2$, Figure 10.4). Nutritionists and food scientists see their research as quite germane to these issues. On the other hand, scientists in both the socioeconomic and the "production" sciences view their research as reflecting consumer issues only to a limited degree. Interestingly, this difference in research focus reflects the difference in the perceived importance of research goals (i.e., health versus productivity) noted in Chapter 9.

Scientists in all disciplines see their research as reflecting alternative approaches to agriculture only to a very limited degree (overall $\bar{x} = 2.6$, Figure 10.5). Entomologists score highest on this issue ($\bar{x} = 3.6$), and food scientists score lowest ($\bar{x} = 1.9$). The relatively low overall level of research effort in this area is of particular concern. It has been suggested that limiting agriculture to a single approach tends to close options and may open agriculture to serious long- and short-run instabilities (National Agricultural Research and Extension Users Advisory Board, 1980). These results underscore the relatively low level of biological and socioeconomic information available on the extent, scope, impact, and import of alternative approaches to agriculture in the United States. It is unlikely that this situation will improve in the near future unless additional funds are made available. The current, relatively low levels of funding in the agricultural sciences often make it necessary for researchers to follow established approaches in order to maintain current levels of agricultural production within the existing system.

Organizational and Extraorganizational Factors

There are minimal differences by organizational and extraorganizational dimensions in the degree to which scientists' research reflects

FIGURE 10.3
Degree to which Research Reflects Energy Issues
by Discipline.

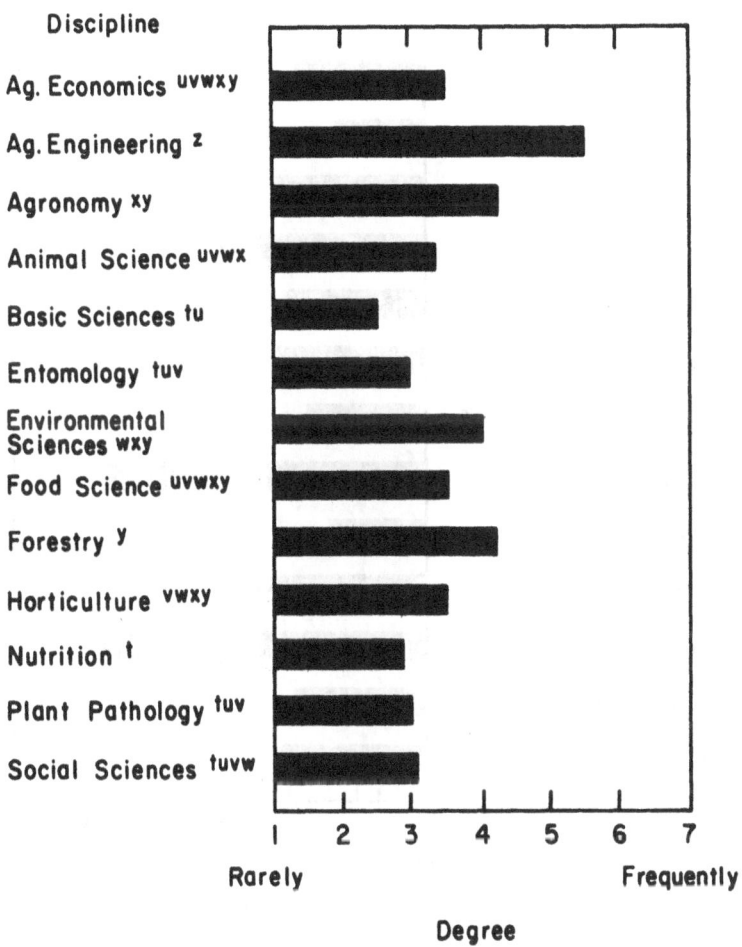

Discipline

Ag. Economics ᵘᵛʷˣʸ

Ag. Engineering ᶻ

Agronomy ˣʸ

Animal Science ᵘᵛʷˣ

Basic Sciences ᵗᵘ

Entomology ᵗᵘᵛ

Environmental
Sciences ʷˣʸ

Food Science ᵘᵛʷˣʸ

Forestry ʸ

Horticulture ᵛʷˣʸ

Nutrition ᵗ

Plant Pathology ᵗᵘᵛ

Social Sciences ᵗᵘᵛʷ

1   2   3   4   5   6   7
Rarely   Frequently

Degree

ᵗ⁻ᶻ Disciplines with different letters are significantly
different p < .01 .

emerging agricultural issues. Between university and federal government scientists, there are no significant differences for four of the
five issues. Scientists employed by the federal government perceive
environmental issues to be significant for their research only slightly
more than those in the experiment stations (government $\bar{x}$ = 4.7,

FIGURE 10.4
Degree to which Research Reflects Consumer Issues
by Discipline .

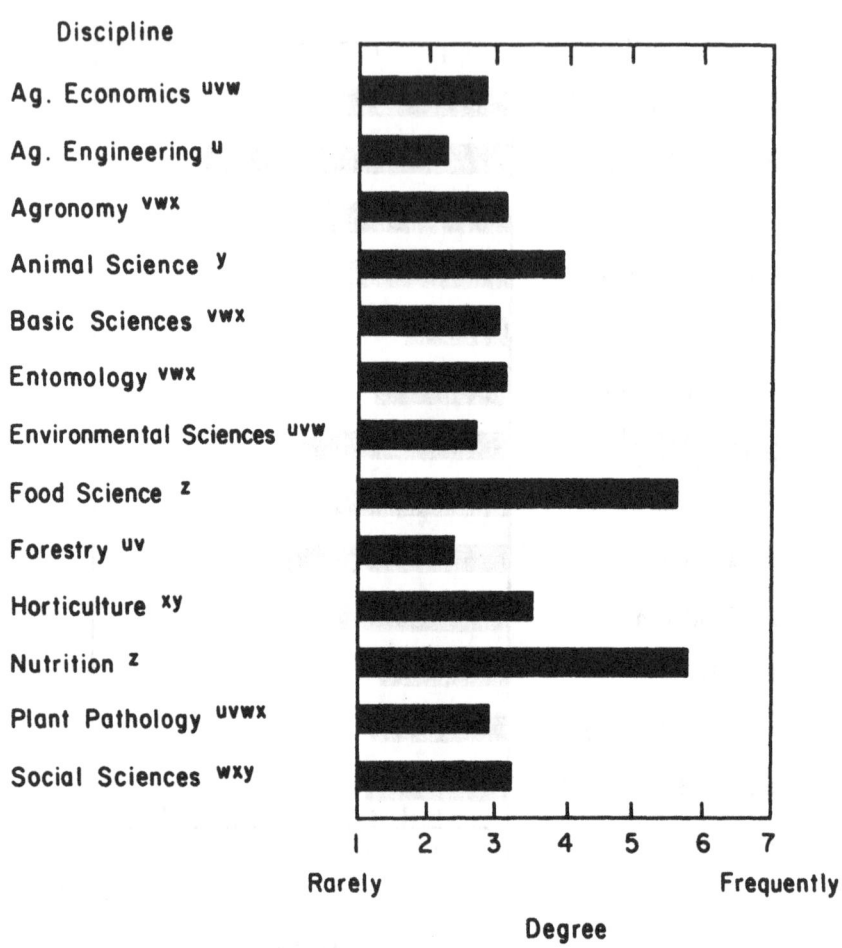

[t-z] Disciplines with different letters are significantly different,
p < .01.

experiment stations $\bar{x} = 4.3$, $p < 0.02$). Institutional prestige, again, appears to make a minimal difference. Only on energy issues (low prestige $\bar{x} = 3.7$, high prestige $\bar{x} = 2.7$; $p < 0.01$) and environmental issues (low prestige $\bar{x} = 4.4$, high prestige $\bar{x} = 3.7$; $p < 0.01$) are there slight, but statistically significant, differences in scientists' perceptions.

The extraorganizational sources of potential influence on scientists are also only moderately related to the degree to which scientists

FIGURE 10.5
Degree to which Research Reflects Alternative Approaches
to Agriculture by Discipline.

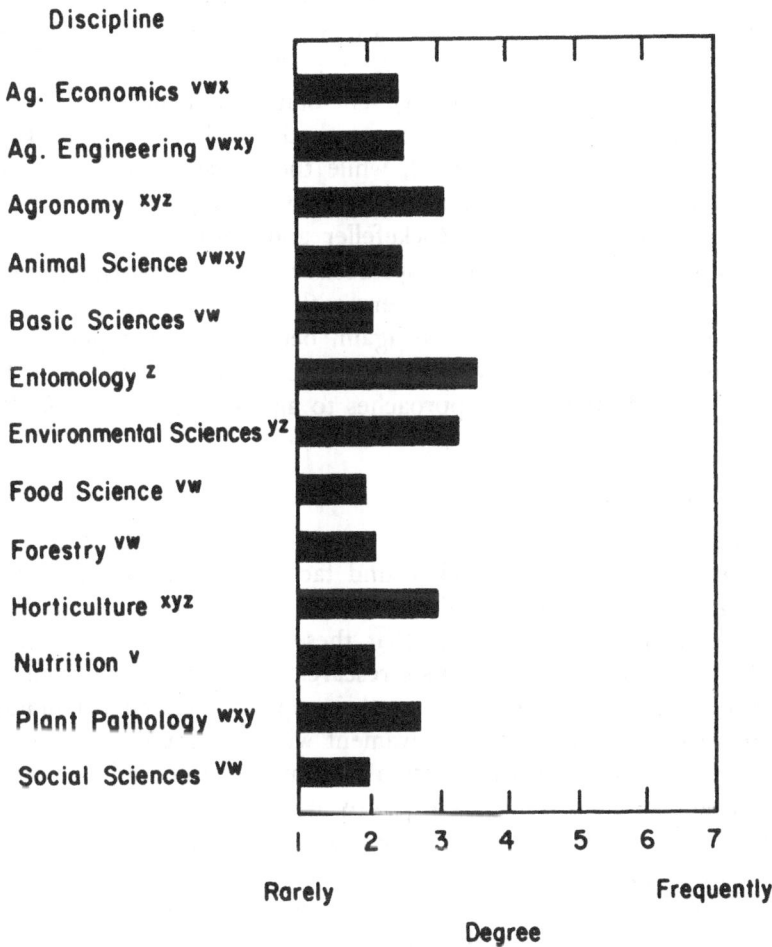

Discipline

Ag. Economics ᵛʷˣ

Ag. Engineering ᵛʷˣʸ

Agronomy ˣʸᶻ

Animal Science ᵛʷˣʸ

Basic Sciences ᵛʷ

Entomology ᶻ

Environmental Sciences ʸᶻ

Food Science ᵛʷ

Forestry ᵛʷ

Horticulture ˣʸᶻ

Nutrition ᵛ

Plant Pathology ʷˣʸ

Social Sciences ᵛʷ

1 2 3 4 5 6 7

Rarely   Frequently

Degree

ᵛ⁻ᶻ Disciplines with different letters are significantly different,
p < .01 .

perceive their research to reflect the five issue groups. The degree
to which scientists' research reflects environmental issues is positively
related to associations with USDA ($r = 0.10$, $p < 0.001$), the Department of the Interior ($r = 0.11$, $p < 0.001$), state government
agencies ($r = 0.12$, $p < 0.001$), and commodity associations ($r =$

0.10, $p < 0.001$). Interestingly, involvement with environmental issues is negatively related to receiving grants from the National Institutes of Health ($r = -0.14$, $p < 0.001$). Perception of one's research as related to world food issues is associated with contacts with some of the same institutions, including the Department of the Interior ($r = 0.12$, $p < 0.001$) and commodity associations ($r = 0.15$, $p < 0.001$), as well as the State Department ($r = 0.13$, $p < 0.001$) and private foundations ($r = 0.08$, $p < 0.001$). The State Department relationship is probably attributable to contact with the U.S. Agency for International Development (USAID), while the private foundation association may be the consequence of international agricultural and food research supported by the Rockefeller and Ford Foundations. The only other significant relationships between these extraorganizational institutions and researchers' concern with the five current issues was the negative relationship, once again, between grant support from NIH and perceived involvement with energy issues ($r = -0.14$, $p < 0.001$) and alternative approaches to agriculture ($r = -0.09$, $p < 0.001$).

Criteria for Problem Choice

The criteria used by scientists to select research problems are affected by a variety of background factors, influential individuals, disciplinary and organizational priorities, and extraorganizational needs and demands. It is also likely that these criteria will be associated with the degree to which scientists' research reflects current agricultural issues. Given the earlier findings that applied research orientations were related to perceived involvement with current issues, it is not surprising to discover that most of the more "practically" oriented and client-oriented criteria are positively related to energy, environmental, food, and consumer issues, and alternative approaches to agriculture (Table 10.2).

Scientists who tend to consider the utility and marketability of their research products, clients' needs, and research organization priorities are also more likely to view their research as reflecting these five issues. Once again, there is the curious lack of relationship between the pursuit of research for its contribution to knowledge and science and the perceived involvement in these five issues. None of the criteria for problem choice that reflect this basic research orientation was associated with perceptions that one's work reflected these topics. This apparent gap between basic research and involvement in current issues facing agriculture, at least in the minds of the scientists, may have important negative consequences for long term approaches to these issues. As a plant physiologist noted: "The theoretical base for new

TABLE 10.2
Correlations Between Scientists' Perceived Research Involvement in
Current Issues and Their Criteria for Problem Choice[a]

Criterion	Alternative Approaches to Agriculture	Consumer Issues	Energy Issues	Environment Issues	World Food Issues
	Current Issues				
Potential creation of new methods, useful devices	.10	.10	.15	.14	.14
Marketability of final product	.16	.13	.22	.13	.16
Publication in farm or industry journals	.18	.11	.18	.08	.08
Publication in bulletins	.14	.03	.16	.09	.05
Importance to society	.16	.15	.15	.15	.25
Demands raised by clientele	.13	.07	.16	.13	.08
Feedback from extension personnel	.16	.08	.18	.14	.14
Client needs as assessed by you	.16	.13	.21	.13	.15
Priorities of the research organization	.13	.13	.14	.13	.18

[a]Pearsonian correlation coefficients (r). Values range from -1.0 (strong negative relationship) to 1.0 (strong positive relationship). A value of 0.0 is interpreted as the complete absence of any relation between the two variables. As the data is ordinal and the coefficients assume interval data, reported results are conservative. If $r > .08$, then $p < .001$.

agricultural technologies is developing in a very slow and spotty way. I think the USDA needs to take more of an intellectual lead and foster more interaction with basic scientists and between basic and applied scientists. Basic scientists' awareness of agricultural research needs also needs to be increased." Another concluded "*Now* is the time to be concerned about 25 years in the future. Apparently . . . basic research with some clear future agricultural applicability is necessary."

Long- and Short-term Research

The importance of long-term as against short-term approaches to research has been an underlying theme throughout this discussion of the emerging agricultural research issues. The time frame for research may be as important for the future of public sector agricultural science as the human resources, capital resources and substance of the research. Generally, the more fundamental concerns of agricultural science, such as biological nitrogen-fixation research and germ plasm alterations, require years of work and long-term planning and funding.

As the current monocultural agricultural system, with its energy and chemical intensive practices, reaches its limits, the possibility of major readjustments in agriculture must be faced. The emerging issues for U.S. agriculture may require fundamentally different practices and technologies that are likely to be developed only through commitment to long-term research. In our study, scientists in a variety of disciplines offered unsolicited observations regarding these issues. A plant bio-chemist commented:

> Too little attention is paid in agriculture (particularly by USDA) to longer range needs and problems, no matter how urgent (or irreversible) they are likely to be. It is dangerous not to consider seriously and *actively* the problems of 5–20 years hence.

An entomologist reiterated this observation:

> In my opinion, formulation of research goals or objectives in entomology are poorly coordinated in the U.S. Communication and coordination between State and Federal research administrators in entomology appears to be poor. Attention is focused on "brush fires" and pressures of the moment with relatively little attention to long-term biological research on population dynamics, behavior, taxonomy.

Numerous scientists ventured possible explanations for the serious deemphasis of long-term research in the system. One viticulturist remarked:

> Research has been drastically curtailed because of present preoccupation with "immediate results." It requires 15 years to produce and adequately test a new grape variety. For any such project nowadays, it is impossible to have support that has the necessary continuity. Despite tremendous outlays for "cancer cure," no immediate results have been forthcoming. One cannot buy scientific breakthroughs. Research can only progress by long-term guarantee of support and less pressure to produce results.

A soil chemist criticized the fickle nature of research funding:

> There has been relatively little communication with USDA related groups. They are not doing basic research in the area I am working

in. It is critical that support of basic research in agriculture have continuity and freedom from bureaucracy that prevents high quality and innovative research in federally controlled programs. The hop on and off bandwagon approach taken by the Congress and administration dissipates energy and funds so that basic understanding is bypassed for collection of data that will be meaningless in five years.

The research funding issue was also raised by a biochemist:

The single greatest problem I face is the need for continuity of research support. Research grants seldom exceed a period of 3 years, and are often not renewed. This causes a general upheaval of the personnel supported on the grant. I have to devote entirely too much attention to proposal writing and report writing.

One forester offered a number of explanations for the lack of long-term research:

Long-term research, such as needed in forestry, is difficult because of the short-term trend in funding projects. [There is] increasing pressure for short discrete problems and quick publication to satisfy needs for publication records. . . . Extremely heavy teaching loads and restricted funding have made the research process more difficult to initiate and carry to completion. Emphasis on journal article publications is influencing the choice of project design, and mode of publication probably is having a negative effect on that developmental/utilization phase of research results and dissemination of results to appropriate public clientele groups.

Additional concern about the role of the publication system in limiting long-term research was expressed by an agricultural economist: "Due to the pressure of rather quickie publication, few young scientists can afford to address the longer range hardcore issues. The academic community has not anticipated issues until they reach the crisis stage." Finally, a botanist perceptively observed:

In weed science, we are using our limited resources of time and money to provide short-term and immediate answers. Thus we spend a disproportionate amount of time on chemical weed control, because it is a way to meet immediate needs. But there is a desperate need for development of long-range strategies for crop protection and pest-management. For example, we need to know how certain crop rotations affect pest populations. We need to evaluate cultural practices for their effects on weeds, pathogens, and insect pests. What are the subtle interactions that favor natural predators of major pest organisms, and how can natural controls be enhanced? Modern farming with its large fields of monocultures and reduced crop rotation appears to favor large-scale pest outbreaks. We cannot go back to diversified small farms, so our response is often an energy-intensive chemical "fix." I believe that with a better understanding of the biology of agriculture, we can reduce many pest problems to acceptable levels depending more on biological and cultural forces. Federal funding for research is already spread very thinly, and even that is dwindling. Agribusiness is not very interested in funding long-range research unless it promises to tie in closely with their markets. The end result is that much important long-range research is not being done.

Coincidental with the demands raised by emerging long-term issues are the more frequent short-term problems that will continue to face this specialized agricultural system. These will include short-term fluctuations in input prices, weather, insects, weeds, and pathogens; availability, quality, and price of foods; and economic returns to producers.

Agricultural research needs to develop appropriate planning and sufficient resources for sustained efforts to meet long-term needs, while retaining the flexibility to respond to urgent short-term problems. A soil scientist offered the following suggestion:

> The most important factor for continued effective agricultural research is that workers are periodically involved with some *real* production or land use problem, so they have a chance to see something that can be the basis for *novel* thinking, then some time to do that work. Secondly, that the problems be chosen *locally, not* in Washington. Thirdly, that research funding be continuous, not closely tied to the popular issues of the day. It need not be large (funds)—but must continue years and years.

Regardless of the particular approach, more careful assessment of long-term priorities, anticipation of short-term problems, and better coordination in determining overall priorities will be required.

Conclusions

In summary, agricultural scientists are addressing a wide range of emerging agricultural issues, but to varying degrees. Our research suggests that, although some issues may be receiving adequate attention, research in other important areas lags behind. In addition, the approaches to these issues may be quite fragmented, as is evidenced by the range of involvement among various disciplines and the relative indifference of some disciplines to particular issues. Moreover, the division between basic and applied scientists may impede the success of long-term research in providing new breakthroughs for these emerging issues. However, it seems unreasonable to expect researchers to address a much broader variety of issues in different ways than they traditionally have, without (1) changes in the training of scientists, (2) modifications of the research system, and (3) substantial increases in both funds and scientific staff. Some possible policy changes for public sector U.S. agricultural research are examined in our concluding chapter.

11
Policy Implications

In this volume we have raised a wide variety of issues relating to the practice of agricultural research in the United States today. These issues have important implications for science, agriculture, and the larger society. In this concluding chapter, we summarize the key findings of this study, examine the changing context for agricultural research and its implications for institutional change, and explore certain epistemological considerations and the role of agricultural research in a democratic society.

Summary of Findings

Historical Development

1. The agricultural sciences developed as part of the expansion of colonial empires and the shift from subsistence to capitalist farming. This origin had the effect of institutionalizing the goal of increased productivity as a central theme in agricultural research. Over the years, increased productivity came to be seen as an end, rather than as a means. Even today, productivity often remains an unchallenged and paramount goal for agricultural research.

2. With the development of an agricultural scientific community at the turn of the century, scientists began to develop their own disciplinary concerns. Like their colleagues in other fields, agricultural scientists became convinced of the efficacy of the scientific method for solving the full range of human problems. As a result, they formed disciplinary associations and began to contribute to increasingly specialized bodies of knowledge. While it appeared at the time that such knowledge would be easily integrated into a coherent whole, the unification of scientific knowledge has proved to be an illusory goal.

Problem Choice

3. Despite the seemingly straightforward character of decisions about problem choice in agricultural research, closer examinination reveals an extraordinarily complex process. In addition to scientific or paradigmatic criteria for problem choice, administrative directives, political commitments, and personal avocations are among the other factors that play a role. These criteria emerge from a complex process of negotiations and often operate together to shape problem choice.

4. Despite official pronouncements and formal structure, scientists report that feedback from extension plays a relatively minor role in research decisions. Furthermore, scientists' personal assessments of client needs are considered substantially more important than either demands raised by clients themselves or feedback from extension staff.

Scientists' Background

5. Large percentages of all age cohorts of agricultural scientists come from farm backgrounds. As the farm population continues to decline, fewer scientists will be drawn from this group. These scientists, in contrast to their colleagues from nonfarm backgrounds, tend to be more concerned with the utility of their research and the needs of client groups such as farmers. Although a farm background is not a prerequisite for agricultural research, an understanding of the everyday problems of farmers and the context within which these problems arise may be important for the creation of relevant research products and for the pursuit of important agricultural issues.

6. Currently, at the public institutions only 4% of agricultural scientists are women. This is a significantly lower percentage than in most other fields of science. Furthermore, most of these women are employed in traditionally female fields (e.g., nutrition, home economics). New opportunities for advancement of women need to be developed.

7. Only 1% of agricultural scientists at public institutions are at the seventeen predominantly black institutions ("1890" schools). These smaller institutions often lack a critical mass of scientists in any one discipline or research area and are generally nonresearch oriented. Clearly, such institutional circumstances severely limit the public agricultural research role of these schools as well as of most black agricultural scientists.

8. Agricultural scientists as a group are slightly older than scientists in related fields. In the coming decades their average age is likely to increase as the agricultural sciences and other sciences experience a period of limited growth and fiscal restraint. Younger scientists,

however, appear more likely to address emerging agricultural issues and may more readily accept new ideas. Therefore, substantial recruitment of new scientists will be important in the coming decades if the agricultural sciences are to retain their vitality. Unfortunately, private industry is now successfully competing for these same scientists, particularly those in certain high technology fields. This raises the possibility that land-grant institutions will be limited in their ability to address emerging agricultural issues and to train new entrants into the high-demand fields because of a lack of trained faculty (see Walsh, 1981).

9. The total career experience for most agricultural scientists tends to be restricted to the land-grant institutions. Only a small subset of these institutions provide career training for the overwhelming majority of agricultural scientists engaged in research. Such a situation is bound to limit scientists' horizons by restricting their exposure to ideas originating outside that system.

10. University scientists see their opportunities to pursue basic research as substantially restricted. Moreover, USDA appears unable to fulfill its basic research role. In both cases, the pressure for immediately applicable results limits work on long-range solutions to agricultural problems. Scientists appear willing and able to do more basic research but are limited by the research system itself.

Scientific Communication

11. Modes of formal scientific communication, such as professional journals, bulletins, and books, are identified by agricultural scientists as essential resources in their research, central for promotion and tenure, and key outlets of their work. In addition, the formal literature scientists read and publish appears to shape the very way in which they pursue their research and choose research agendas. However, this formal communication has become highly specialized and increasingly dominated by domestic disciplinary journals. Furthermore, scientists who rely more heavily on professional journals are more discipline and science oriented and less concerned with products and clients, whereas those who view bulletins as more important deemphasize disciplinary criteria for their work and stress utilitarian aspects of their research. Consequently, the varieties of formal communication upon which scientists rely may move them in very different and potentially contradictory directions.

12. Informal communication about research among agricultural scientists and with other system participants is relatively infrequent and limited primarily to contact with scientists in one's own department. Even within this group, researchers communicate less often than once

a week. Informal contact with other groups, such as scientists in other disciplines and institutions and clients, occurs less often than once a month. Our results and numerous studies note the importance of such informal communication to research agendas and to scientific advance.

13. Despite infrequent contact with other scientists, staff, and clients, agricultural scientists indicate that a number of these people are influential in a variety of research decisions. One of the most important sources of influence is one's colleagues at another institution. With decreasing funds available for travel to other campuses or professional meetings, scientific advance may be hindered.

14. The influence of basic scientists appears minimal in the agricultural sciences. Although agricultural basic scientists perceive other disciplines as the most important beneficiaries of their research, other agricultural scientists infrequently read basic science journals, rarely speak to basic scientists about their research, and perceive basic scientists as uniformly noninfluential for nearly all research decisions.

Disciplines and Organizations

15. Agricultural research is increasingly fragmented along disciplinary lines. Scientists typically receive all or most of their education within the same discipline. They rarely subscribe or publish outside disciplinary lines. As each discipline develops a somewhat different vocabulary, cross-disciplinary communication is restricted.

16. The prestige of land-grant institutions and their agricultural orientation appears to be negatively related. Scientists at high prestige institutions seem to have relatively little interest in applied agricultural questions and client needs, while those at low prestige institutions are quite concerned with such issues. This suggests that the prestige of universities within the land-grant system tends to be a function of the disciplinary standing of their respective departments.

17. Scientists are agreed that the reward system tends to overemphasize the publication of large numbers of articles in scientific journals. Clearly, such a reward system encourages the unnecessary proliferation of scientific papers and may encourage a higher priority for disciplinary problems over problems of societal concern. (This is not to suggest an inherent conflict between the two, but few mechanisms exist to insure that disciplinary objectives will be consonant with societal needs.) The reward system inadequately recognizes unusual and innovative research that is long-term in scope or unpublishable in disciplinary journals.

18. Research grants, contracts, and consulting simultaneously pull scientists in myriad directions. Many scientists with grants from government agencies appear far less concerned with agricultural issues

than their colleagues who are not grant recipients. On the other hand, scientists with commodity association and private corporation grants frequently appear to make corporate goals their own. In some fields consulting levels are so high as to raise serious questions about the ability of scientists to perform adequately their research functions. The current fiscal climate is likely to magnify such problems, particularly as universities and, to a lesser extent, USDA agencies search for new ways to compensate for declining budgets.

Beneficiaries, Goals, and Issues

19. Scientists tend to see themselves as serving virtually all groups of beneficiaries equally. However, careful studies of beneficiaries suggest that certain groups are served to a greater degree than others. For example, the tendency of agricultural research products to be capital intensive tends to benefit those individuals with ready capital to invest, i.e., large farmers and agribusinesses.

20. Scientists overwhelmingly emphasize the creation of disciplinary knowledge and the increase of agricultural productivity as the most important goals for agricultural research. Many other goals, such as human nutrition, improving rural levels of living, and improving communities, tend to be relegated to one or two disciplines, In general, scientists appear relatively unaware of the value of the research of other agricultural disciplines.

21. There is a surprising lack of concern for contemporary agricultural issues among agricultural scientists, particularly among basic scientists and those with a basic research orientation. This raises some important questions about the probability of adequately addressing these issues, as the most promising approaches to them are likely to require substantial long-term, basic science inputs.

22. When concern is expressed for contemporary agricultural issues, it is often limited to a few selected disciplines. Environmental issues, the most frequently considered research issues, are most likely to be reflected in the work of environmental scientists and entomologists. Energy issues are addressed mainly by agricultural engineers, agronomists, and foresters. Few biologists or social scientists have dealt with these issues. Consumer issues appear to be the domain of food scientists and nutritionists. Finally, there is a virtual absence of researchers examining alternative approaches to agricultural production.

23. The time frame for research may be as important for the problems facing public sector agricultural research as is the substance of the research. The emerging questions facing the research community may require fundamentally different practices and technologies that are likely to be developed only through commitment to long-term

research. As a consequence of funding practices, reward structures, immediate needs and demands of clientele, and organizational priorities, current research appears limited to short-term projects.

Agricultural research has helped the United States to produce an abundant food supply, but it now faces a number of new challenges as it enters its second century. None of these challenges is insurmountable if effective action is taken. Such action requires an understanding of the changing social, political, and economic context for research. It also requires a response to this context on at least two levels: organizational and epistemological.

The Changing Context for Research

A public image of the scientist as an observer who carefully records for posterity detailed observations of some aspect of nature still exists. Indeed, as little as 50 years ago, there was still much truth in this Cartesian view of science. At that time a scientist might be employed in some nonscientific occupation during the day, while working each evening in his small laboratory or study. In a few selected cases, the scientist might enjoy the support of a patron.

Congruent with this view of the scientist as a lone observer was the traditional view of the state. From the days of Adam Smith to the present, it has been asserted that the state has as its primary function the maintenance of order and stability so that economic—and scientific—life would be unhindered. Implicit in this view was the notion that the state transcended the interests of any special group and was, thus, essentially, a neutral institution.

In recent years the Cartesian image of the lone scientist working through the night has been shattered. It has been replaced by a Baconian corporate science in which large numbers of scientists work together within the walls of a complex administrative structure. At the same time the role and scope of the state have been radically altered. Regulatory agencies, tariff barriers, subsidies, direct government grants, and burgeoning bureaucracy have become the rule rather than the exception. To a great extent, the state has become the servant of a variety of overlapping, and sometimes conflicting, interest groups. Indeed, as federal support for ailing industries and the enormous corporate influence on public policy suggest, the very distinction between public and private has been blurred.

Science, and, particularly, agricultural science within the United States, has been dependent upon the generosity of the government. This largesse has contributed substantially to the vitality of U.S. agriculture. At the same time, such support has had a number of limitations. As we noted in the first chapter, over the years research administrators have courted various interest groups at both state and national levels in order to insure continued funding of research. In so doing, perhaps they have become the servants of power, as Hightower (1973), Berry (1977), and others have suggested. If such is the case, scientists have been unhappy with that role. Many critics, however, fail to realize that without this active cultivation of support, public agricultural research would cease to exist. In fact, this is precisely what has happened in a number of Latin American countries (Trigo and Piñeiro, 1981) as well as in U.S. poultry research (Savage, 1975).

In an effort to cultivate farmer support, administrators have encouraged researchers to work on topics of immediate concern to farmers. However, researchers and administrators alike appear to have failed to notice that farmers' demands are conditioned by the current organization of production and marketing. By taking that socioeconomic organization as a given, agricultural research has assumed that the most effective way to ameliorate farm problems is through increasing purchases of manufactured inputs. This, in turn, has condemned farmers to the "product market 'treadmill'" (Ruttan, 1980). It has also indirectly "subsidized" agribusiness by creating an increasing demand for its goods and services. Rather than making farmers more independent, it has limited their options.

Some researchers, seeking relief from the constraints imposed by the political context of this system, have looked to certain granting agencies for support (e.g., NIH, NSF). Scientists successful in obtaining grants, however, have often found that such support was available only if one severed one's ties with the problems of agriculture and rural life. This has tended to fragment and to divide the agricultural research system.

Agricultural researchers and research administrators face serious funding problems. The exigencies of organizational maintenance have demanded that research administrators become entrepreneurs (Rosenberg, 1976). But traditional sources of support have begun, within the last decade, to decline. Particularly at the national level, the broad farm groups have been increasingly overshadowed by a myriad of commodity associations. These powerful interest groups have led to a greater fragmentation in the overall support for research. And, at the same time, formula funds for research—particularly federal funds—measured in constant dollars, have stagnated. Administrators, however reluctantly, have felt compelled to encourage scientists to seek external funds. Finally, demands for accountability have made administrators hesitant to provide funds for high risk research projects.

A host of new interest groups have emerged in the last decade.

Organic farmers, long considered to be outside both the political and the scientific mainstream, have succeeded in establishing an organic farming unit at USDA's Beltsville Agricultural Research Center (see Youngberg, 1978). Backyard gardening, long the province of a dedicated few, has become increasingly significant in produce markets. As a result, the once dormant state and national horticultural societies have been steadily gaining in membership. Small and medium-sized farmers have begun to engage in a variety of direct marketing arrangements, ranging from farmers' markets to pick-your-own operations. The consumer movement has also made known its interests in nutrition and food safety. Of particular concern to consumer advocates have been questions of proper diet and of the safety of various food additives. Similarly, the environmental movement has demanded additional research on farm runoff, pollution from animal wastes, and soil erosion. Both the environmental and the consumer movements played a role in the recent establishment of several environmental toxicology departments at agricultural colleges. Farm labor organizing, begun in the 1960s, has become an important force in farm politics in a number of states. The suit brought on behalf of farm workers against the University of California represents a growing understanding among farm workers of the role of agricultural research policy in changing the labor needs of U.S. agriculture.

The emergence of these and other groups on the agricultural scene has changed markedly the demands made upon agricultural research. (A recent U.S. GAO report made a similar observation about extension services [1981].) Obviously, these various groups seek potentially conflicting objectives. As a result, many researchers and administrators have either embraced only cautiously or studiously avoided these new interests. Such persons have been content to retreat to the older, more established organizations with which they feel most comfortable. Only in a few scattered instances have research administrators actively sought support from these new groups.

The emergence of these new interest groups creates both a danger and an opportunity for the future of public agricultural research. On the one hand, squabbling between the new and the more traditional groups may well lead to a reduction in overall support and even more fragmentation in agricultural research. On the other hand, the very unsettled character of the current scene may permit the emergence of more effective leadership from the research community.

Ironically, it was the very success of agricultural research that helped to change farming from a way of life for half the population into an occupational group with its set of special interests. Research has often encouraged the differentiation of farmers into commodity specialists, the increased separation of farming (tilling the soil) from agriculture (the food system), and the increased industrialization of input production and food processing. In so doing, research has played a major role in sweeping away the foundations of legitimacy that made possible broad public support for agricultural research in the first place. Only by broadening its support base through the cultivation of multiple, and often conflicting, support groups will public agricultural research play its part in solving the problems that lie ahead. Such a reorientation of the research system will be at best difficult. Yet, without it, it is possible that public research will gradually cease to be supported, and key problems outside the domain of commodity interests will remain unresearched. The development of such a broadened base of support will require the development of new institutional structures that can effectively communicate with and define the needs of groups without a commodity orientation. Such a reorientation will require that new kinds of educational, research, and extension programs be added to those currently existing. It will also require a rethinking of the epistemological base for agricultural research.

Suggestions for Institutional Change

We do not here propose a sweeping and expensive reorganization of the U.S. agricultural research system. Restructuring research institutions is a difficult task. Often efforts to restructure institutions result in little more than additional layers of bureaucracy. It is true that public agricultural research appears to be underfunded and would probably benefit from a change in agenda coupled with increased research budgets, but there is no reason to belabor this point. Rather, in light of both the findings of our study and the current problems facing U.S. agriculture, we suggest the following modest changes:

1. There is a need for an organization that would speak for the public agricultural research community at the national level, provide a national forum for leadership, and develop guidelines for research and evaluation. At one time the special needs of agriculture were well understood within the National Association of State Universities and Land-Grant Colleges. Today agricultural research has been eclipsed by the enormous growth in other fields. A revitalized Experiment Station Committee on Organization and Policy or perhaps a new national organization representing teaching, research, and extension interests in agriculture (including those in USDA) could meet these needs.

2. More interdisciplinary communication and research within the agricultural sciences and between agricultural and nonagricultural

disciplines is sorely needed. This can be accomplished on several grounds at once:

a. Graduate students could be encouraged to take minors outside of their disciplines. For example, some agronomists might be encouraged to minor in animal nutrition.
b. Faculty could be encouraged to take sabbatical and postdoctoral leaves at non-land-grant institutions and in disciplines other than their own.
c. Agricultural disciplinary societies could establish annual awards for the best interdisciplinary research.
d. Disciplinary societies might consider frequently inviting distinguished scientists from other disciplines to present state-of-the-art papers and otherwise participate at their annual meetings.
e. More joint appointments across departmental lines (in both government and the universities) could be developed.
f. Administrators should consider the establishment of several multidisciplinary, multicommodity research institutes. Such institutes might employ an approach similar to the farming systems approach adopted by the International Agricultural Research Centers. In fact, they might be jointly funded with USAID, thereby formally linking domestic and international research interests. Koppel (1981) and Flinn (1980) noted the declining capabilities of the land-grant system in the international arena. However, this linkage is particularly important in light of the development of a significant number of new approaches to agricultural research in other nations.

The new Board on Agriculture of the National Research Council may be a constructive first step. However, as the director of the Michigan Experiment Station (Wittwer, 1980) has noted, much of public food and agricultural research policy is made outside of the agricultural research community. Furthermore, "what has been done with respect to the Food and Agriculture Act of 1977 in establishing the Joint Council and the Users Committee is to keep the participant occupied rather than really influencing policy" (Wittwer, personal correspondence, 1982).

3. There is a need for a high quality interdisciplinary U.S. agricultural journal. Currently, nearly all publication opportunities in agricultural research lie in disciplinary journals. The basic biological sciences have journals, such as *Science*, that provide an outlet for superior papers in the field, a forum for interdisciplinary research, and a record of news of interest to and about the scientific community. No such journal

currently exists in the agricultural sciences, although the *Agricultural Science Review* at one time partly filled that role. Such a journal, perhaps jointly supported by the various disciplinary societies, would provide a new, prestigious outlet for research outside traditional disciplinary boundaries.

4. In hiring decisions, scientists and research administrators should take into account an applicant's breadth of knowledge and experience as well as his or her understanding of a particular special area. In addition, administrators need actively to seek qualified women and minority members for research positions. By encouraging greater breadth and diversity among agricultural scientists we may achieve the goals of greater flexibility in meeting changing needs, increased interdisciplinary interaction, and, perhaps, greater creativity.

5. The current reward systems in agricultural research institutions need to be reorganized so as to encourage scientists to pursue more high risk research and long-term research. The emphasis within universities upon the publication of scientific journal articles stems in part from the diminished role of colleges of agriculture within those universities. University administrators often seek standardized measures for evaluating the performance of all faculty and ignore the special roles of agricultural research. Better systems for promoting and rewarding high quality research that addresses national goals need to be developed. In this process, the panacea of easily quantified performance evaluations must be avoided.

6. The focus of institutional advisory groups needs to be expanded to encompass a broader range of issues and client concerns. Most colleges of agriculture and federal research agencies have multiple advisory groups, oriented almost entirely along commodity lines. Thus, there are committees for beef cattle, cotton, grapes, and so on. Only rarely have advisory groups effectively represented farmers who have chosen, for reasons of capital shortage, risk, or personal preference, not to specialize. This commodity orientation not only gives specialists an edge, but also discourages research that considers the entire farm enterprise rather than a single commodity. By creating *farmer* (as opposed to commodity) advisory groups, administrators can indirectly encourage interdisciplinary research and deliver services to a currently underserved population.

A similar case may be made for the development of new kinds of nonfarm advisory groups made up of consumers, environmentalists, backyard gardeners, and those persons farming by nontraditional methods. Representatives of such groups would significantly broaden the agricultural research agenda and provide a climate more receptive to scientific innovations.

7. There is an urgent need to develop a truly national agricultural research policy and to integrate it into a coherent national agricultural policy. Neither a coherent research policy nor an agricultural policy currently exists. The present processes used to set research priorities largely serve to aggregate what researches are already doing. Setting new directions will require:

a. The creation of new institutional forms by which the full range of constituents for agricultural research can be represented, and

b. the establishment of *several* multidisciplinary units in which policy research can be conducted. Some of the larger, land-grant universities appear to have the resources as well as the academic freedom necessary to carry out this latter task.

8. The development of policy also requires the ability to assess the results of policies once they are implemented and, eventually, the ability to develop forecasting techniques. Friedland and Kappel (1979) urged the development of social impact statements for agricultural research. Such a statement differs from an environmental impact statement in that it "creates no basis for immediate action but is stored for later evaluation although there is a clear understanding that, at some stage when the methodology of social impact assessment improves, the statements will become the basis for decision making" (1979:28). Such statements would be brief and would be prepared by a special "Predictive and Evaluative Methodology Unit" on each agricultural campus.

We believe these suggestions form the basis for significant institutional change in agricultural research. They avoid both the tendency to respond immediately to producer demands and the opposite tendency to divorce agricultural research from agriculture by embracing "pure" science. Nevertheless, institutional changes are insufficient without concomitant changes in what constitutes knowledge in the agricultural sciences.

Epistemological Considerations

At first glance it may seem inappropriate to raise epistemological questions in a book of this kind. However, the institutional changes suggested above are probably insufficient to surmount the problems facing U.S. agriculture and the agricultural research community today.

As was noted in Chapter 1, agricultural research in the United States began during a period in which science was overtly and solidly reductionist. Knowledge of the natural and social world was to be obtained by subdividing that world into finer and finer pieces. In addition, the idea of an overarching world view that served to frame research questions was rejected outright. Adherence to the scientific method would insure that each research result would be a building block in the cathedral of science.

Over the last several decades, this position has increasingly come under attack. Thomas Kuhn (1970) emphasized the importance of alternative world views or paradigms in the formulation of scientific research programs (see Busch and Lacy, 1981). Similarly, work by von Bertalanffy (1968) emphasized the limits of a reductionist approach to science and suggested the appropriateness of a systems perspective as an alternative. (For a general systems approach applied directly to agricultural research, see Dillon [1976].)

The existence of a reductionist epistemology in the agricultural sciences is well known. Its influence has been felt in many ways. For instance, plant taxonomist Edgar Anderson complained of the fragmentation traceable to a reductionist approach:

> It is difficult to find men or groups of men who can and will work effectively in several of these fields at once. Problems which fall straight across departmental and divisional lines run into administrative red tape. A whole series of coherent, fundamental questions are neglected because they do not fall clearly within the domain of any single discipline (1967:108).

Anderson went on to emphasize the importance of having a broad understanding as well as technical expertise (1967).

More recently, horticulturist W. Grierson asserted that "another fetter on the imaginative young scientists, is the use of statistics as an enforcement agent for iron-bound orthodoxy rather than as a sharp tool to aid research" (1980:228–229). While mathematical and statistical formulations permit the use of a common language across disciplines, they must necessarily exclude those aspects of the world not amenable to mathematical formulation.

This same reductionist stance encourages certain kinds of mundane research while discouraging higher risk, but imaginative, approaches to research. Grierson provided an excellent description, for example, of why it is easier to test the impact of a chemical on a particular plant than to do imaginative research. He noted that chemicals can be routinely tested by utilizing well-established methods; imaginative research by definition must challenge the status quo.

Agronomist D. G. Hanway lamented the reductionism and over-specialization common in this field:

> Up to the present time we have not really developed comprehensive, integrated, multidisciplinary research programs that deal with improvement of crop production systems as systems. . . . In the U.S. most individual components are studied independently of others. Recommendations and educational efforts are carried out by specialists each traveling his own way and telling his own story. The farmer is confronted with making a system out of all the diverse information that comes his way. I'm sure experiment stations have not often assembled all the components of the systems they recommend to see how they function together (1978:5).

Hanway also noted that most crop breeding programs have emphasized yield potential while ignoring nutritional value (1978).

An example of the practical problems that may result from a reductionist approach to research is the case of processing tomatoes reported by the National Commission on Productivity (1973:18). Between 1960 and 1970, the net weight of tomatoes purchased by processors in California increased by 10%. This was the result of research designed to increase the weight of tomatoes produced per acre. However, all of the increased weight was due to additional water content—water that had to be removed in processing! The net result was higher fuel bills for processors and higher prices for consumers.

Still another example of the unidimensionality of the reductionist approach was the Council on Agricultural Science and Technology (CAST) panel on organic farming. (For discussions of the role of CAST see Anthan [1979], Henig [1979], and Webster [1979].) Members of the panel were informed by the chairman at their first meeting that the purpose of the panel was not to evaluate the effectiveness, limitations, and applicability of organic farming, but rather to show that it was thoroughly and wholly inappropriate. A public outcry from scientists who, though not proponents of organic farming, felt it deserved more careful scientific consideration eventually led to a satisfactory conclusion of the incident. Nevertheless, it pointed out the unidimensional understanding of the nature of knowledge that appears all too frequently in current research.

However, in recent years developments have occurred that suggest the emergence of new approaches to agricultural research. At a minimum, such approaches are multidisciplinary, holistic, and diverse. Let us briefly consider some examples of these new developments.

Plant breeder Charles Francis (1980), in a paper read to the American

Seed Trade Association, reflected recently upon future directions for plant breeding. Francis proposed the development of mechanized intercropping systems, as well as whole new approaches to agronomic problems, including the development of perennial cereal crops, the use of permanent sod cultures, and the creation of multiple purpose crops.

In their recent textbook on plant pathology, Horsfall and Cowling (1977) devoted chapters to "societal contraints on management" and "the sociology of plant pathology." While many issues remain unaddressed, these authors' attempts both to examine how plant pathology fits into the larger social picture and to understand something of the dynamics of plant pathology as a science represent an important step forward.

The new Center on Agricultural and Environmental Values established at Cornell University represents an important attempt to re-establish a link between the humanities—particularly philosophy—and the natural sciences in agriculture. Though it is too soon to assess the work of the Center, it holds the promise of making a significant contribution to interdisciplinary understanding.

Without exaggerating, one can argue that Integrated Pest Management (IPM) has significantly changed the direction of research in economic entomology. In the past much research in this area could be characterized as the testing of various chemicals for effectiveness in pest eradication. At a minimum, IPM has replaced the notion of pest eradication with the more ecologically sound notion of managing the size of insect populations. In addition, it has shifted the center of entomological research away from chemical control measures to the development of control strategies employing a combination of biological, mechanical, and chemical means. Acceptance of IPM strategies has also made it far more acceptable, if not imperative, for entomologists to collaborate directly with their colleagues in agronomy and horticulture in an effort to develop better systems of management. The recently conceptualized Integrated Reproductive Management offers prospects similar to those of IPM. However, this framework will be a more difficult one to build in light of the sharper division between plant and animal scientists.

Within the last few years the image of the future projected by agricultural engineers has shifted markedly. In a special bicentennial issue of *Agricultural Engineering*, Frank Lanham (1976) described a totally engineered agriculture of the future. By 1979 the official historian of the ASAE, Robert E. Stewart, was already pointing out that such technology might be both socially undesirable and impossible (1979). A more recent issue of the journal showed the significant changes over the last few years in engineers' understandings of what is possible and desirable (Christianson et al., 1980).

Until recently, nearly all research conducted in the public sector focused upon the development of capital-intensive agricultural systems employing purchased inputs. The development of a research unit on organic farming within USDA represents a significant change in policy. It also paves the way for research on other alternative approaches to agriculture.

Farming systems research (FSR), developed over the last decade largely out of the experience of the International Agricultural Research Centers, has begun to surface in the United States (e.g., Hansen et al., 1981). As a recent appraisal of farming systems research suggested, "the primary aim of the FSR approach is to increase the productivity of the farming system in the context of the entire range of private and societal goals, given the constraints and potentials of the existing farming systems" (Gilbert, Norman, and Winch, 1980:2). Effective farming systems research requires not only the creation of multidisciplinary research teams, but also the reconstruction of the very epistemology employed in research. As Gilbert, Norman, and Winch stated, "priorities for research reflect the holistic perspective of the whole farm/rural household and the natural and human environment" (1980:3). Thus, farming systems research is concerned not with the optimization or maximization of production of a particular commodity, but rather with the optimization of the farming system as a whole. This tends to refocus the emphasis in agricultural research away from disciplinary and commodity concerns toward complex interactions among and between people, crops, soil, and livestock. Were it not for the general predilections of ecologists to study only systems untouched by human hands, farming systems research would clearly be called a branch of ecology. Regardless of the label, farming systems research offers significant advantages for both farmers and researchers.

Finally, the recent work by soil scientist Yvon Chatelin is significant in that it marks a new level of awareness on the part of agricultural scientists. Chatelin's *An Epistemology of the Soil Sciences* (1979) showed how conceptual problems within soil science are traceable to the divergent epistemological assumptions of nominalism and realism. Chatelin proposed that reductionist approaches be replaced by a dialectical soil science in which soils are viewed as dynamic totalities rather than as static fragments. Such an approach does not exclude the use of symbols, formalizations, and mathematical relations, nor does it grant them primordial importance. It provides a comprehensive view of the soil and then places it in the larger context of the biosphere.

Each of the examples cited above has the potential to alter significantly the current understanding and practice of agricultural research so that it may more effectively take into account the larger social and natural world. To do so, the institutional changes described above and others will be needed. Whether there is the leadership and will to change remains to be seen.

Science, Agriculture, and Democracy

In this day and age, talk about the relationship between agriculture and democracy sounds somehow quaint. It conjures up romantic, Jeffersonian images of a nation of yeoman farmers forever safe from the evils of manufacturing and urban life—independent and able to participate as equals in the governance of the community. The Morrill Act of 1862 espoused similar ideals. The land grants were to be used to provide "democracy's colleges," places where farmers and "mechanics" could go to receive an education at once liberal and technical. The colleges were to prepare students not only for political democracy but for a democracy that would "permeate all aspects of the work and life of the people" (Butterfield, 1918:45).

Today, yeoman farmers exist only as images in the minds of romantic writers. In addition, in recent years democracy has been taken "more or less for granted, as if democracy were self-perpetuating rather than something which has to be re-won through self-discipline by each generation for itself" (Hutchison, 1944:25). Instead, universities have focused increasingly upon the technical knowledge they wish to impart to students. Scientists, as both teachers and students, have become increasingly specialized in both their scientific and their social lives.

Some years ago, Dean C. B. Hutchison of the University of California at Berkeley explained the problem as follows:

Specialization is the easy road in the intellectual journey of life. . . . Specialization contributes to the development of a stratified and compartmentalized society which, more than any other single factor, is making it difficult for democracy to succeed. Most specialists—most technical graduates—though scientific in their own fields, are often misinformed, frequently dogmatic and sometimes even biased and prejudiced in other fields. . . . Graduates of our agricultural, engineering, and home economics curricula need to be more than good technicians—more than competent, skilled, professional men and women. They need to be educated individuals capable of thinking and acting intelligently as citizens in a free society (1944:26–27).

In short, specialization without breadth in both science and society is antithetical to democracy. It separates scientist from scientist, farmer from farmer. It inhibits the development of new approaches because it limits the options known by a given individual to be available. Further, it restricts popular participation because it divides society into ever narrower segments, each "expert" in his or her little corner of the world.

On the other hand, we should note that democracy flourishes when there are broad, common understandings shared by all. This is true of both the agricultural scientific community and the larger society in which it is embedded. In short, agricultural research and the land-grant/USDA system can be a force for the maintenance and improvement of democracy by encouraging greater participation in research decision making both within the scientific community and with the various publics it serves.

Some years ago a noted agricultural scientist wrote: "In our investigations we still stress too much the goal of increased productivity as our great task. We still have too much faith in knowledge of the physical and biological facts and principles as all sufficing. There should be searchings of heart as to our policies and programs. Are they adequate to the needs of the new epoch?" We believe that this question, raised by Kenyon Butterfield in 1917, (1918:54) is still valid today.

References

Abelson, Philip H.
  1980   "Scientific communication." *Science* 209:60–62.

Agricultural Engineering
  1978   "ASAE adopts complete goals program." *Agricultural Engineering* 59 (April):48–49.
  1979   "The issues in American agriculture today. . . ." *Agricultural Engineering* 60 (April):8–15, 23.

Allen, T. J.
  1969   "Information needs and uses." *Annual Review of Information Science and Technology* 4:3–29.
  1970   "Communication networks in R&D laboratories." *R&D Management* 1:14–21.

Anderson, Edgar
  1967   *Plants, Man and Life.* Berkeley: University of California Press.

Anthan, George
  1979   "Scientists resign in protest of 'bias'."*Des Moines Register* (January 21):1F–2F.

Arndt, Thomas M.; Dana Dalrymple; and Vernon Ruttan (eds.)
  1977   *Resource Allocation and Productivity in National and International Agricultural Research.* Minneapolis: University of Minnesota Press.

Arndt, Thomas M., and Vernon Ruttan
  1977   "Valuing the productivity of agricultural research." Pp. 3–25 in Thomas Arndt, Dana Dalrymple, and Vernon Ruttan (eds.), *Resource Allocation and Productivity in National and International Agricultural Research.* Minneapolis: University of Minnesota Press.

Babb, E. M.
  1979   "Some causes of structural changes in U. S. agriculture." Pp. 51–60 in *Structure Issues of American Agriculture.* USDA, Economics, Statistics, and Cooperatives Service, AER 438. Washington, D.C.

Bailey, Liberty Hyde
1910   "The better preparation of men for college and station work." Pp.
[1909]   25–32 in *Proceedings of the 23rd Annual Convention of the Association of American Agricultural Colleges and Experiment Stations.* Portland.
1920   *Field and Purpose of the American Association of Agricultural Legislation.* Secretary of the AAAL, Bulletin No. 5. Madison: University of Wisconsin.

Barham, W. S.
1977   "Industry-institutional cooperation and coordination in horticulture: A questionnaire." *HortScience* 12:3 (June):208–10.

Barre, H. W.; L. E. Call; J. C. Kendall; C. A. Moore; E. W. Allen; and J. T. Jardine
1928   "Report of the experiment station committee on organization and policy." Pp. 203–5 in *Proceedings of the 42nd Annual Convention of the National Association of Land-Grant Colleges and Universities.* Washington, D.C.

Bauman, Howard E.
1979   "Changes in industrial research require changes in graduate education." *Food Technology* 33 (12):30–31.

Beardsley, Edward H.
1969   *Harry L. Russell and Agricultural Science in Wisconsin.* Madison: University of Wisconsin Press.

Bergland, Robert, and Susan Sechler
1981   *Time to Choose: Summary Report on the Structure of Agriculture.* Washington, D.C.: USDA.

Bernal, J. D.
1948   *Report on the Royal Society Scientific Information Conference.* London: Royal Society.

Berry, Wendell
1977   *The Unsettling of America: Culture and Agriculture.* Totawa, New Jersey: Sierra Club Books.

Beyer, Janice M.
1978   "Editorial policies and practices among leading journals in four scientific fields." *Sociological Quarterly* 19:68–88.

Bonnen, James T.
1962   "Some observations on the organizational nature of a great technological payoff." *Journal of Farm Economics* 44:1279–94.
1967   "The crises in the traditional roles of agricultural institutions." Pp. 48–62 in Vernon W. Ruttan, Arley D. Waldo, and James P. Houck (eds.), *Agricultural Policy in an Affluent Society.* New York: W. W. Norton and Company.

Boss, Andrew
  1927 "Agriculture as a business compared to agriculture as a mode of life." Pp. 141–49 in *Proceedings of the 41st Annual Convention of the Association of Land-Grant Colleges and Universities.* Chicago.

Bredahl, Maury E.; W. Keith Bryant; and Vernon W. Ruttan
  1980 "Behavior and productivity implications of institutional and project funding of research." *American Journal of Agricultural Economics* 62:371–83.

Breimyer, Harold F.
  1973 "Discussion." *American Journal of Agricultural Economics* 55:993–96.

Brennan, Patrick W.
  1975 "Informational flow in American agricultural literature." *Quarterly Bulletin of the International Association of Agricultural Librarians and Documentalists* 20:86–93.

Broad, William J.
  1981 "Nutrition research: End of an empire." *Science* 213:518–20.

Brockway, Lucille H.
  1979 *Science and Colonial Expansion.* New York: Academic Press.

Brunner, Edmund de Schweinitz
  1957 *The Growth of a Science: A Half-century of Rural Sociological Research in the United States.* New York: Harper.

Buchanan, R. E.
  1938 "Should it be the function of the land-grant college to implement all federal programs in agriculture?" Pp. 108–18 in *Proceedings of the 52nd Annual Convention of the Association of Land-Grant Colleges and Universities.* Chicago.

Busch, Lawrence
  1978 "On understanding understanding: Two views of communication." *Rural Sociology* 43:450–74.
  1980 "Structure and negotiation in the agricultural sciences." *Rural Sociology* 45:26–48.

Busch, Lawrence, and William B. Lacy
  1981 "Sources of influence on problem choice in the agricultural sciences, the New Atlantis revisited." Pp. 113–28 in Lawrence Busch (ed.), *Science and Agricultural Development.* Totawa, New Jersey: Allanheld, Osmun.

Busch, Lawrence; William B. Lacy; and Carolyn Sachs
  1980 *Research Policy and Process in the Agricultural Sciences: Some Results From a National Study.* RS-66. Lexington: University of Kentucky, Department of Sociology.
  1983 "Perceived criteria for research problem choice in the agricultural sciences." *Social Forces* (in press).

Busch, Lawrence, and Carolyn Sachs
  1981   "The agricultural sciences and the modern world system." Pp. 131–56 in Lawrence Busch (ed.), *Science and Agricultural Development*. Totawa, New Jersey: Allanheld, Osmun.

Butterfield, Kenyon
  1907   *Chapters in Rural Progress*. Chicago: University of Chicago Press.
  1918   "The Morrill Act institutions and the new epoch." Pp. 43–49 in
  [1917] *Proceedings of the 31st Annual Convention of the Association of American Agricultural Colleges and Experiment Stations*. Washington, D.C.

Campbell, Christiana McFayden
  1962   *The Farm Bureau and the New Deal*. Urbana: University of Illinois Press.

Carlson, C. W.; D. J. Bray; and G. J. Mountney
  1977   "Scientist year and publication expenditures for U.S. poultry research in 1974 and 1975." *Poultry Science* 56:1960–67.

Carson, Rachel
  1962   *Silent Spring*. Boston: Houghton Mifflin.

Carter, Harold O.
  1976   "U.S. agricultural and food policy: A case of conflicting goals." *Economic and Social Issues* (April-May). Davis: University of California, Cooperative Extension.

Cartwright, Richard Wayne
  1971   "Research Management in a Department of Agricultural Economics." Ph.D. dissertation, Purdue University.

Chatelin, Yvon
  1979   *Une Epistemologie des Sciences du Sol*. Memoires O.R.S.T.O.M. No. 88. Paris: Organisation de Recherche Scientifique et Technique Outre Mer.

Cheney, H. B.
  1974   "Roles of the agronomist." *Agronomy Journal* 66: 1–4.

Christianson, Leslie L.; R. George; M. D. Shanklin; N. F. Meador; and G. L. Hahn
  1980   "Farmstead facilities and arrangements for 2025." *Agricultural Engineering* 61:17–18.

Chubin, Daryl E.
  1980a "Competence is not enough." [Review] *Contemporary Sociology* 9:204–7.
  1980b "Reply to the Coles." [Letter] *Contemporary Sociology* 9:604–5.

Clinton, L. A.; W. W. Long; G. I. Christie; R. S. Wilson; Thomas Bradlee; and K. L. Hatch
  1921   "Report of the committee on extension organization and policy." Pp. 223–25 in *Proceedings of the 35th Annual Convention of the Association of State Universities and Land-Grant Colleges*. New Orleans.

Cole, Jonathan R., and Stephen Cole
1973 *Social Stratification in Science.* Chicago: University of Chicago Press.

Cole, Stephen; Jonathan R. Cole; and Gary A. Simon
1981 "Chance and consensus in peer review." *Science* 214:881–86.

Cole, Stephen; Leonard Rubin; and Jonathan R. Cole
1978 *Peer Review in the National Science Foundation: Phase One of a National Study.* Washington: National Academy of Sciences.

Coleman, A. Lee
1979 "How do agriculture college students differ by major?" Lexington: University of Kentucky, Agricultural Experiment Station, Department of Sociology, mimeographed.

Colman, Gould P., and Jerry D. Stockdale
1976 "Area development through agricultural innovation: New York's sugar beet fiasco." Rural Sociological Society, Monograph No. 4. Morgantown: West Virginia University.

Commission on Country Life
1911 *Report to the Commission on Country Life.* New York: Sturgis and
[1909] Walton.

Commoner, Barry
1972 *The Closing Circle, Nature, Man, and Technology.* New York: Alfred A Knopf.

Convention of Friends of Agricultural Education
1967 *An Early View of the Land-Grant Colleges.* Urbana: University of Illinois
[1871] Press.

Cook, Robert E.
1977 "Poultry science 1976." *Poultry Science* 56:3–7.

Cooper, Thomas; A. Atkinson; L. N. Duncan; F. D. Farrell; C. A. Lory; H. A. Morgan; H. W. Mumford; G. F. Warren; and F. W. Peck
1927 "Report of the special committee as to the agricultural situation." Pp. 88–114 in *Proceedings of the 41st Convention of the American Association of Land-Grant Colleges and Universities.* Chicago.

Coulter, John Lee
1928 "Organized labor and capital and unorganized agriculture." Pp. 108–16 in *Proceedings of the 42nd Annual Convention of the Association of Land-Grant Colleges and Universities.* Washington, D.C.

Cox, R. S.
1976 "Letters." *Phytopathology News* 10:3.

Crane, Diana
1969 "Fashion in science: Does it exist?" *Social Problems* 16:433–41.
1972 *Invisible Colleges: Diffusion of Knowledge in Scientific Communities.* Chicago: University of Chicago Press.

Culliton, Barbara J.
1979 "President Carter addresses National Academy." *Science* 204:480–81.

Dana, Samuel T.
1950   "Forest research." Pp. 316–39 in Robert K. Winters (ed.), *Fifty Years of Forestry in the U.S.A.* Washington: Society of American Foresters.

Danbom, David B.
1979   *The Resisted Revolution.* Ames: Iowa State University Press.

Darwin, Charles
1966   *On the Origin of Species: A Facsimile of the First Edition.* Cambridge,
[1859] Massachusetts: Harvard University Press.

Davenport, Eugene
1919   "Wanted: A national agricultural policy." Address of the President of the Association of American Agricultural Colleges and Experiment Stations. Urbana: University of Illinois.

Davidson, J. Brownlee
1927   "Agricultural engineering developments and economic adjustments."
[1926] Pp. 177–85 in *Proceedings of the 40th Annual Convention of the Association of Land-Grant Colleges and Universities.* Washington, D.C.

Day, Boysie E.
1978   "The morality of agronomy." Pp. 19–27 in J. W. Pendleton (ed.), *Agronomy in Today's Society.* American Society of Agronomy Special Publication No. 33. Madison.

Day, Edmund E.
1943   "The lost battalion." Pp. 24–32 in *Proceedings of the 57th Annual Convention of the Association of Land-Grant Colleges and Universities.* Chicago.

Dillman, D. A.; James A. Christenson; Edwin H. Carpenter; and Ralph M. Brooks
1974   "Increasing mail questionnaire response: A four-state comparison." *American Sociological Review* 39:744–56.

Dillon, John L.
1976   "The economics of systems research." *Agricultural Systems* 1:5–22.

Duniway, C. A.
1915   "Economic science in agricultural and mechanical colleges." Pp. 94–96 in *Proceedings of the 29th Annual Convention of the Association of American Agricultural Colleges and Experiment Stations.* Berkeley.

Dupree, A. Hunter
1957   *Science in the Federal Government.* Cambridge, Massachusetts: Belknap Press.

Ebbinghaus, James H.
1978   "Farm machinery of the future." Pp. 103–16 in *Proceedings and Minutes of the 27th Annual Meeting of the Agricultural Research Institute.* Washington, D.C.

Edge, David O., and Michael J. Mulkay
1976   *Astronomy Transformed: The Emergence of Radio Astronomy in Britain.*
New York: Wiley-Interscience.

Erdman, Henry Ernest
1929   "Who gets the benefit of improvement in agriculture?" *Journal of Farm Economics* 11:24–43.

Evenson, Robert E., and Yoav Kislev
1975   *Agricultural Research and Productivity.* New Haven: Yale University Press.

Ewing, John A.
1975   "The role of agricultural economics in a changing environment." *Southern Journal of Agricultural Economics* 7:1–4.

Faux, Geoffrey
1973   "Reclaiming America: A proposal for California." *Working Papers* 1:31–42.

Feigenbaum, Edward D.
1980   "Solving the 'working-for-the-sponsor syndrome' in externally-funded research organizations." *Journal of the Society for Research Administrators* 12:41–43.

Feyerabend, Paul
1975   *Against Method: Outline of an Anarchistic Theory of Knowledge.* London: New Left Books.
1978   *Science in a Free Society.* London: New Left Books.

Fifield, Willard
1954   "The role of federal and state institutions in the Agricultural Research Institute." Pp. 16–20 in *Proceedings of the 3rd Annual Meeting of the Agricultural Research Institute.* Washington, D.C.

Finegold, Kenneth
1982   "From agrarianism to adjustment: The political origins of new deal agricultural policy." *Politics and Society* 11:1–27.

Fishel, Walter (ed.)
1971   *Resource Allocation in Agricultural Research.* Minneapolis: University of Minnesota Press.

Fiske, Emmett P.
1978   "From communities to commodities." Paper presented at the Annual Meetings of the Rural Sociological Society. San Francisco.
1979   "The College and Its Constituency: Rural and Community Development at the University of California, 1875–1978." Ph.D. dissertation, University of California, Davis.

Flinn, William L.
1980   "The graying of rural sociologists in international development and related social class concerns." *Newsline* 8:32–35.

Forman, Paul; John Heilbron; and Spence Weart
  1975   *Physics circa 1900: Personnel, funding, and productivity of the academic establishment.* Historical Studies in the Physical Sciences, Vol. 5. Princeton: Princeton University Press.

Francis, Charles A.
  1980   "Developing hybrids of corn and sorghum for future cropping systems." Paper presented at 35th Annual Corn and Sorghum Industry Research Conference, American Seed Trade Association. Chicago.

Friedland, William H.
  1979   "Who killed rural sociology? A case study in knowledge production." Paper presented at the Annual Meetings of the Rural Sociological Society, Boston.

Friedland, William H., and Amy Barton
  1975   *Destalking the Wily Tomato: A Case Study in Social Consequences in California Agricultural Research.* Davis: University of California, Department of Behavioral Sciences, Research Monograph No. 15.

Friedland, William H.; Amy E. Barton; and Robert J. Thomas
  1978   *Manufacturing Green Gold: The Conditions and Social Consequences of Lettuce Harvest Mechanization.* Davis: University of California, Department of Applied Behavioral Sciences.

Friedland, William H., and Tim Kappel
  1979   *Production or Perish: Changing the Inequalities of Agricultural Research Priorities.* Santa Cruz: Project on Social Impact Assessment and Values, University of California.

Frome, Michael
  1971   *The Forest Service.* New York: Praeger.

Fry, Bernard, and H. S. White
  1975   *Economics and Interaction of the Publisher-Library Relationships in the Production and Use of Scholarly and Research Journals.* Bloomington: Indiana University, Graduate Library School.

Fujimoto, Isao, and Emmett Fiske
  1975   "What research gets done at a land-grant college: Internal factors at work." Davis: University of California, Department of Applied Behavioral Science, mimeographed.

Fujimoto, Isao, and William Kopper
  1975   "Outside influences on what research gets done at a land-grant school: Impact of marketing orders." Paper presented at the Annual Meetings of the Rural Sociological Society. San Francisco.

Ganz, Carole
  1976   "The role of scientific communication in the process of technological innovation." *ASLIB Proceedings* 28:385–91.

Gardner, V. R.; P. V. Cardon; and S. W. Fletcher
   1935   "Report of the special committee on publication of research." Pp.
          189–95 in *Proceedings of the 49th Annual Convention of the Association
          of Land-Grant Colleges and Universities.* Washington, D.C.

Garfield, Eugene
   1979   *Citation Indexing—Its Theory and Application in Science, Technology and
          Humanities.* New York: John Wiley.

Garvey, William D., and S. D. Gottfredson
   1977   "Scientific communication as an interactive social process." *Inter-
          national Forum on Information and Documentation* 2:9–16.

Garvey, William D., and B. C. Griffith
   1967   "Scientific communication as a social system." *Science* 157:1011–16.

Garvey, William D.; Nan Lin; Carnot E. Nelson; and Kazuo Tomita
   1972   "Research studies in patterns of scientific communication: II. The
          role of the national meeting in scientific and technical communi-
          cation." *Information Storage and Retrieval: Theory and Practice* 8:159–69.

Garvey, William D.; Nan Lin; and Kazuo Tomita
   1972   "Research studies in patterns of scientific communication: III. In-
          formation exchange process associated with the production of journal
          articles." *Information Storage and Retrieval: Theory and Practice* 8:207–21.

Gerstl, J. E., and S. P. Hutton
   1966   *Engineers: The Anatomy of a Profession.* New York: Tavistock Publi-
          cations.

Gieryn, Thomas F.
   1978   "Problem retention and problem change in science." *Sociological
          Inquiry* 48:96–115.

Gilbert, E. H.; David W. Norman; and F. E. Winch
   1980   *Farming Systems Research: A Critical Appraisal.* Rural Development
          Paper No. 6. East Lansing: Michigan State University, Department
          of Agricultural Economics.

Glaser, Barney G., and Anselm L. Strauss
   1967   *The Discovery of Grounded Theory: Strategies for Qualitative Research.*
          Hawthorne, New York: Aldine Publishing Company.

Goldschmidt, Walter
   1978   *As You Sow.* Montclair, New Jersey: Allanheld, Osmun.

Goss, Kevin F., and Richard D. Rodefeld
   1977   "Consequences of Mechanization in U.S. Agriculture." Paper pre-
          sented at the Annual Meetings of the Rural Sociological Society.
          Madison.

Greenberg, D. S.
   1966   "'Bootlegging': It holds a firm place in conduct of research." *Science*
          153:848–49.

Grierson, W.
1980   "The enforced conservatism of young horticultural scientists."
         HortScience 15:228–29.

Guterman, C.E.F.
1944   "The Cornell Research Foundation." Pp. 134–36 in *Proceedings of
         the 58th Annual Convention of the Association of Land-Grant Colleges
         and Universities.* Chicago.

Haberer, Joseph
1969   *Politics and the Community of Science.* New York: Van Nostrand
         Reinhold.

Habermas, Jurgen
1970   *Towards a Rational Society.* Boston: Beacon Press.

Hadwiger, Don F.
1975   "The green revolution: Some new perspectives." *Change* 7:36–62.
1982   *The Politics of Agricultural Research.* Lincoln: University of Nebraska
         Press.

Hagstrom, Warren O.
1965   *The Scientific Community.* New York: Basic Books.

Hamilton, T. S.
1954   "For the state experiment stations." Pp. 70–75 in *Proceedings of the
         3rd Annual Meeting of the Agricultural Research Institute.* Washington,
         D.C.

Hansen, Art; D. Griffith; J. Butler; S. Powers; E. Gilbert; R. Lauriault; M.
Downie; J. Dean; and S. Russo
1981   *Farming Systems of Alachua County, Florida: An Overview with Special
         Attention to Low Resource Farmers.* Gainesville: University of Florida,
         Institute of Food and Agricultural Sciences, Center for Community
         and Rural Development.

Hanway, D. G.
1978   "Agricultural experiment stations and the variety protection act,
         Part 2." *Crops and Soils* 30:5–7.

Hardin, Charles
1946   "The bureau of agricultural economics under fire: A study in
         valuation conflicts." *Journal of Farm Economics* 28:635–68.
1947   "Programmatic research and agricultural policy." *Journal of Farm
         Economics* 29:359–82.
1955   *Freedom in Agricultural Education.* Chicago: University of Chicago
         Press.

Harmon, Lindsey R.
1965   *Profiles of Ph.D.'s in the Sciences.* Washington, D.C.: National Academy
         of Sciences.
1978   *A Century of Doctorates: Data Analyses of Growth and Change.* Wash-
         ington, D.C.: National Academy of Sciences.

Harper, Alfred E.
1979   "Bad advice to Congress: A critique." *Nutrition Today* 14:22–26.

Hatch, William Henry
1886   "Agricultural experiment stations." *Report of the Committee on Agriculture, U.S. House of Representatives.* Washington, D.C: USGPO.

Havelock, Ronald G.
1971   *Planning for Innovation Through Dissemination and Utilization of Knowledge.* Ann Arbor: University of Michigan.

Hays, Samuel P.
1959   *Conservation and the Gospel of Efficiency.* Cambridge, Massachusetts: Harvard University Press.

Heady, Earl O.
1949   "Basic economic and welfare aspects of farm technological advance." *Journal of Farm Economics* 31:293–316.
1971   "Welfare implications of agricultural research." Pp. 121–36 in Walter L. Fishel (ed.), *Resource Allocation in Agricultural Research.* Minneapolis: University of Minnesota Press.
1972   "Allocations of colleges and economists." *American Journal of Agricultural Economics* 54:934–44.

Henig, Robin Marantz
1979   "Agriculture's strange bedfellows: CAST-industry tie raises credibility concerns." *BioScience* 29:9–12, 59.

Hightower, Jim
1973   *Hard Tomatoes, Hard Times.* Cambridge, Massachusetts: Schenckman.

Hilgard, E. W.
1885   "The functions of experiment stations." *Science* 5:23.

Hooks, Gregory M., and William L. Flinn
1981   "The country life commission and early rural sociology." *The Rural Sociologist* 1:95–100.

Horsfall, James G., and Ellis B. Cowling
1977   "Societal constraints on management." Pp. 103–19 in James G. Horsfall and Ellis B. Cowling (eds.), *Plant Disease: An Advanced Treatise.* New York: Academic Press.

Hudgins, Garven F. (ed.)
1981   *Proceedings of the 95th Annual Meeting of the National Association of State Universities and Land-Grant Colleges.* Washington, D.C.

Hull, David F.; Peter D. Tessner; and Arthur M. Diamond
1978   "Planck's principle: Do younger scientists accept new scientific ideas with greater alacrity than older scientists?" *Science* 202:717–23.

Hutchison, C. B.
1944   "The liberal education of the 'Industrial Classes'." Pp. 24–42 in *Proceedings of the Annual Convention of the Association of Land-Grant Colleges and Universities.* Chicago.

Huxley, L. (ed.)
1901   *Life and Letters of Thomas Henry Huxley.* 2 vols. New York: Appleton.

Jimenez, Mario G.
1964   "Needed research on the flow of information among scientists and other personnel connected with research activities." Pp. 28–43 in D. T. Myren (ed.), *Proceedings of the First InterAmerican Symposium on the Role of Communictions in Agricultural Development.* Mexico City.

Jones, C. R.
1917   "Scientific management as applied to the farm, home, and manu-
[1916] facturing plants." Pp. 108–15 in *Proceedings of the 30th Annual Convention of the Association of American Agricultural Colleges and Experiment Stations.* Washington, D.C.

Jordan, W. H.
1907   "The authority of science." Pp. 60–66 in *Proceedings of the 21st Annual Convention of the Association of American Agricultural Colleges and Experiment Stations.* Lansing, Michigan.

Kaldor, Donald R.
1971   "Social returns to research and the objectives of public research." Pp. 62–79 in Walter L. Fishel (ed.), *Resource Allocation in Agricultural Research.* Minneapolis: University of Minnesota Press.

Kellogg, Charles E., and David C. Knapp
1966   *The College of Agriculture: Science in the Public Service.* New York: McGraw-Hill.

Kelly, Patrick; Melvin Kranzberg; Stanley R. Carpenter; and Frederick A. Rossini
1977   *The Flow of Scientific and Technical Information in the Innovation Process: An Analytical Study.* Atlanta: Department of Social Sciences, Georgia Technical University.

Kenney, Martin; Frederick H. Buttel; J. Tadlock Cowan; and Jack Kloppenburg, Jr.
1982   "Genetic engineering and agriculture." Paper presented at the workshop on Biotechnology, Agriculture, and the Public Interest, Lexington, Kentucky.

Kerr, Howard W.
1980   *A Survey of Current and Expected Research Needs of Small Farms in the Northwestern Region.* USDA, Science and Education Administration, ARR-NE-9. Washington, D.C.

Kidd, Charles V.
1959   "Basic research—description versus definition." *Science* 129:368–71.

King, D. W.; D. D. McDonald; N. K. Roderer; and B. L. Wood
1976   *Statistical Indicators of Scientific and Technical Communication 1960–1980.* Vol. 1. *A Summary Report for the National Science Foundation.* Rockville, Maryland: Center for Quantitative Sciences.

Kirkendall, Richard S.
1966 *Social Scientists and Farm Politics in the Age of Roosevelt*. Columbia: University of Missouri Press.

Klein, Arthur J.
1930 *Survey of Land-Grant Colleges and Universities*. 2 vols. Washington, D.C.: U.S. Department of Interior, Office of Education.

Knoblauch, H. C.; E. M. Law; and W. P. Meyer
1962 *State Agricultural Experiment Stations: A History of Research Policy and Procedure*. USDA, Miscellaneous Publication No. 904. Washington, D.C.

Knorr, Karin D., and Dietrich W. Knorr
1978 "From scenes to scripts: On the relationship between laboratory research and published papers in science." Paper presented at the Annual Meetings of the American Sociological Association. San Francisco.

Koppel, Bruce
1981 "Report of a workshop on linkages between the international and domestic functions of the land-grant system." Honolulu: East-West Center.

Krumland, R. B.; E. E. Will; and G. A. Gorry
1979 "Scientific publications of a medical-school faculty." *Journal of Medical Education* 54:876–84.

Kuhn, Thomas S.
1970 *The Structure of Scientific Revolutions*. 2nd ed. Chicago: University of Chicago Press.
1977 *The Essential Tension: Selected Studies in Scientific Tradition and Change*. Chicago: University of Chicago Press.

Kumlein, W. F.
1927 "Responsibilities of the land-grant colleges in teaching agriculture as a way of life." Pp. 149–55 in *Proceedings of the 41st Annual Convention of the Association of Land-Grant Colleges and Universities*. Chicago.

Lacy, William B., and Lawrence Busch
1982a "The institutional and professional context for rural sociology." Pp. 404–13 in Don A. Dillman and Daryl J. Hobbs (eds.), *Rural Society in the United States: Issues for the 1980s*. Boulder, Colorado: Westview Press.
1982b "Guardians of science: Journals and journal editors in the agricultural sciences." *Rural Sociology* 47 (3): 429–48.

Lacy, William B.; Lawrence Busch; and Carolyn Sachs
1980 "Perceived criteria for research problem choice in the agricultural sciences." Paper presented at the Fifth World Congress for Rural Sociology. Mexico City.

Lacy, William B.; Kenneth Pigg; and Lawrence Busch
1980   "Clients, colleagues, and colleges: Perceived influences on extension agents." *Rural Sociology* 45:469–82.

Ladd, Everett Carl, Jr., and Seymour Martin Lipset
1979   "How professors rated faculties in 19 fields." *The Chronicle of Higher Education* (January 15), 6–7.

Land, George T. Lock
1973   *Grow or Die.* New York: Delta.

Lang, Serge
1981   *The File.* New York: Springer-Verlag.

Lanham, Frank B.
1976   "Heritage and horizons—How it began." *Agricultural Engineering* 57 (December):19–34.

Latour, Bruno, and Steve Woolgar
1979   *Laboratory Life: The Social Construction of Scientific Facts.* Beverly Hills: Sage Publications.

Lee, Linda
1978   *A Perspective on Cropland Availability.* USDA, Economics and Statistics Service, AER 406. Washington, D.C.

Leiss, William
1972   *The Domination of Nature.* New York: George Braziller.

Levins, Richard
1973   "Fundamental and applied research in agriculture." *Science* 181: 523–24.

Lewin, J.
1972   "A quantitative and qualitative case-study analysis of scientific productivity in agricultural research." *Israel Journal of Agricultural Research* 22:129–39.

Lewontin, Richard C.
1982   "Agricultural research and the penetration of capital." *Science for the People* 14:12–17.

Lindsey, Duncan
1978   *The Scientific Publication System in Social Science.* San Francisco: Jossey-Bass.

Lower, William C., and Carey L. Quarles
1975   "Student enrollment survey of poultry science majors in land-grant institutions." *Poultry Science* 54:955–58.

Lu, Yao-Chi
1979   "Technological change and structure." Pp. 121–27 in *Structure Issues of American Agriculture.* USDA, Economics, Statistics, and Cooperatives Service, AER 438. Washington, D.C.

Luria, S. E.
1973 "On research styles and allied matters." *Daedalus* 102:75–84.

Lynn, K. C.
1969 "A quantitative comparison of conventional information compression techniques in dental literature." *American Documentation* 20:149–51.

Madden, J. Patrick, and E. J. Parterheimer
1972 "Evidence of economies and diseconomies of farm size." Pp. 91–107 in Earl O. Heady and Gordon A. Ball (eds.), *Size, Structure, and Future of Farms.* Ames: Iowa State University Press.

Magrabi, Frances M.
1969 "A systems approach to the organization of agricultural research." *Agricultural Science Review* 7 (2):26–32.

Mahoney, Michael J.
1979 "Psychology of the scientists: An evaluative review." *Social Studies of Science* 9:349–75.

Mainzer, Lewis C.
1958 "Science democratized: Advisory committees on research." *Public Administration Review* 18:314–23.

Maritain, Jacques
1944 *The Dream of Descartes.* New York: Philosophical Library.

Marshall, Eliot
1980 "Bergland opposed on farm machine policy." *Science* 208:578–80.

Martin, M. W., and R. L. Ackoff
1963 "The dissemination and use of recorded scientific information." *Management Science* 9:322–36.

Mayberry, B. D. (ed.)
1977 *Development of Research at Historically Black Land-Grant Institutions.* Tuskegee Institute: Association of Research Coordinators, Land Grant 1890 Colleges and Universities.

McCalla, Alex
1973 "Public sector research and education and the agribusiness complex: Unholy alliance or socially beneficial partnerships?" *American Journal of Agricultural Economics* 55:1001.
1977 "Politics of the U.S. Agricultural Research Establishment." Paper prepared for the Agricultural Policy Symposium. Washington, D.C.

McConnell, Grant
1953 *The Decline of Agrarian Democracy.* Berkeley: University of California Press.

Meadows, A. J.
1974 *Communication in Science.* London: Butterworth and Company.

Medawar, P. B.
1967 *The Art of The Soluble.* London: Methuen.

Meier, Kenneth John
    1978   "Client representation in USDA bureaus: Causes and consequences."
            Policy Studies Journal 6:484-89.

Menzel, H.; L. Lieberman; and J. Dulchin
    1960   *Review of Studies in the Flow of Information Among Scientists.* New York:
            Bureau of Applied Social Research, Columbia University.

Merrill, E. D.
    1928   "Cooperation with organized agricultural producers." Pp. 101–4 in
            *Proceedings of the 42nd Annual Convention of the Association of Land-
            Grant Colleges and Universities.* Washington, D.C.

Merton, Robert K.
    1970   *Science, Technology and Society in Seventeenth Century England.* New
    [1938] York: Harper and Row.
    1973   *The Sociology of Science.* Chicago: University of Chicago Press.

Millar, Gordon H.
    1969   "Objectives of industrial research." Pp. 39–48 in *Research with a
            Mission.* American Society of Agronomy Special Publication No. 14.
            Madison.

Miller, Thomas A.; G. E. Rodewald; and Robert McElroy
    1981   *Economies of Size in Major Field Crop Farming Regions of the United
            States.* Washington, D.C.: USDA, Economics and Statistics Service.

Mills, C. Wright
    1959   *The Sociological Imagination.* New York: Grove Press.

Minckler, Leon S.
    1976   "Directions of forest research in America." *Journal of Forestry*
            74:212–16.

Mitroff, Ian
    1974   *The Subjective Side of Science.* New York: Elsevier.

Moore, Ernest G.
    1967   *The Agricultural Research Service.* New York: Praeger.

Morrow, G. E.
    1895   "President's Address." Pp. 25–31 in *Proceedings of the Eighth Annual
    [1894] Convention of the Association of American Agricultural Colleges and
            Experiment Stations.* Washington, D.C.

Moseman, Albert H.
    1971   "Research systems." Pp. 139–48 in Melvin G. Blase (ed.), *Institutions
            in Agricultural Development.* Ames: Iowa State University Press.

Mulkay, Michael J.
    1977   "Sociology of the scientific research community." Pp. 93–148 in
            Ina Spiegel-Rosing and Derek DeSolla Price (eds.), *Science, Technology
            and Society.* London: Sage.

Mumford, F. B.
1926   "The function of research in developing a national agriculture
[1925] policy." Pp. 188–93 in *Proceedings of the 39th Annual Convention of
        the Association of Land-Grant Colleges.* Chicago.

Myers, John A.
1889   "Station organization and methods." Pp. 99–101 in *Proceedings of
        the 2nd Annual Convention of the Association of American Agricultural
        Colleges and Experiment Stations.* Knoxville, Tennessee.

Nagi, Saad, and Ronald G. Corwin
1972   "The research enterprise: An overview." Pp. 1–27 in Saad Nagi
        and Ronald G. Corwin (eds.), *The Social Contexts of Research.* New
        York: John Wiley.

NASULGC and USDA
1966   *A National Program of Research for Agriculture. Report of a study sponsored
        jointly by USDA and NASULGC.* Washington, D.C.: USGPO.

National Academy of Sciences
1979   *Research Excellence Through the Year 2000: The Importance of Maintaining
        a Flow of New Faculty into Academic Research.* Washington, D.C.:
        National Academy of Sciences.

National Agricultural Research and Extension Users Advisory Board
1980   *Report to the President and Congress.* Washington, D.C.: USGPO.

National Association of Conservation Districts
1980   *Soil Degradation: Effects on Agricultural Productivity.* National Agri-
        cultural Lands Study Interim Report No. 4. Washington, D.C.:
        National Association of Conservation Districts.

National Commission on Productivity
1973   *Productivity in the Food Industry: Preliminary Study of Problems and
        Opportunities.* Washington, D.C.: USGPO.

National Research Council
1972   *Report of the Committee on Research Advisory to the U.S. Department of
        Agriculture [Pound Report].* National Technical Information Service,
        PB 213 338. Washington, D.C.

National Research Council, Commission on Human Resources
1976   *Doctoral Scientists and Engineers in the United States: 1975 Profile.*
        Washington, D.C.: National Academy of Sciences.

National Research Council, Office of Scientific Personnel
1971   *Mobility of Ph.D.'s Before and After the Doctorate.* Washington, D.C.:
        National Academy of Sciences.

National Science Board
1981   *Science Indicators 1980: Report of the National Science Board 1981.*
        Washington, D.C.: National Science Foundation.

Nicholson, Heather Johnston
1977   "Autonomy and accountability in basic research." *Minerva* 15:32–61.

Olpin, A. R.
   1944   "The Ohio State University Research Foundation." Pp. 125–31 in *Proceedings of the 58th Annual Convention of the Association of Land-Grant Colleges and Universities.* Chicago.

O'Neal, Edward A.
   1936   "Unified agriculture." Pp. 47–52 in *Proceedings of the 50th Annual Convention of the Association of Land-Grant Colleges and Universities.* Houston.

Paarlberg, Don
   1981   "The land-grant colleges and the structure issue." *American Journal of Agricultural Economics* 63:129–34.

Paisley, William J.
   1966   *The Flow of (Behavioral) Science Information: A Review of the Research Literature.* Palo Alto: Institute for Communication Research, Stanford University.

Patterson, Fred L. (ed.)
   1976   *Agronomic Research for Food.* American Society of Agronomy Special Publication No. 26. Madison.

Pearl, Raymond
   1915   "The publication of the results of investigations made in experiment stations in technical scientific journals including the Journal of Agricultural Research." Pp. 186–92 in *Proceedings of the 29th Convention of the Association of American Agricultural Colleges and Experiment Stations.* Washington, D.C.

Pearse, Andrew
   1980   *Seeds of Plenty, Seeds of Want.* Oxford: Oxford University Press.

Pelz, Donald C., and Frank M. Andrews
   1966   *Scientists in Organizations.* New York: John Wiley.
   1976   *Scientists in Organizations.* Revised Ed. Ann Arbor: Institute for Social Research.

Penn, J. B.
   1979   "The structure of agriculture: An overview of the issues." Pp. 2–23 in *Structure Issues of American Agriculture.* USDA, Economics and Statistics Service, AER-438. Washington, D.C.

Perelman, Michael
   1982   "Who does the present system of agriculture serve?" Paper presented to Conference on Agriculture, Change and Human Values. The University of Florida, Gainesville, Florida.

Perkins, John H.
   1977   "Insecticides and the environment: Intervention and turmoil in applied entomology." Paper presented at the 15th International Congress for the History of Science. Edinburgh.

Piven, Frances Fox, and Richard A. Cloward
1971 *Regulating The Poor.* New York: Random House.

Planck, M.
1950 *Scientific Autobiography and Other Papers.* London: Williams and Norgate.

Potter, A. A.
1923 "Tendencies and problems in land-grant institutions." Pp. 27–32 in *Proceedings of the 36th Annual Convention of the Association of Land-Grant Colleges.* Washington, D.C.

Price, Derek DeSolla
1963 *Little Science, Big Science.* New York: Columbia University Press.

Price, Derek DeSolla, and Suha Gursey
1975 "Some statistical results for the numbers of authors in the states of the United States and the nations of the world." Pp. 26–34 in *Who is Publishing in Science.* Philadelphia: Institute for Scientific Information.

Radnitzky, Gerard
1973 *Contemporary Schools of Metascience.* Chicago: Henry Regnery Company.

Ravetz, Jerome R.
1971 *Scientific Knowledge and Its Social Problems.* New York: Oxford University Press.

Richter, Maurice N., Jr.
1972 *Science as a Cultural Process.* Cambridge, Massachusetts: Schenkman Publishing Company.

Rogers, Everett M., and F. Floyd Shoemaker
1971 *Communication of Innovations, A Cross-Cultural Approach.* New York: The Free Press, Macmillan Publishing Company, Inc.

Rose, Hilary, and Steven Rose
1976 "The problematic inheritance: Marx and Engels on the natural sciences." Pp. 1–13 in Hilary Rose and Steven Rose (eds.), *The Political Economy of Science.* New York: Holmes and Meier.

Rosenberg, Charles E.
1964 "The Adams Act: Politics and the cause of scientific research." *Agricultural History* 38:3–12.
1971 "Science, technology, and economic growth: The case of the agricultural experiment station scientist, 1875–1914." *Agricultural History* 45:1–20.
1976 *No Other Gods: On Science and American Thought.* Baltimore: Johns Hopkins University Press.

Rosenbloom, R., and F. Wolek
   1970   *Technology and Information Transfer.* Cambridge, Massachusetts: Harvard University Press.

Rosenthal, Robert
   1966   *Experimenter Effects on Behavioral Research.* New York: Appleton-Century-Crofts.

Rossiter, Margaret W.
   1979   "The organization of the agricultural sciences." Pp. 211–48 in Alessandra Oleson and John Voss (eds.), *The Organization of Knowledge in Modern America.* Baltimore: Johns Hopkins University Press.

Runyan, Jack L.; John C. Bouma; and Gerald E. Peck
   1975   "Food distribution research priorities to allow major improvements in total systems productivity by 1985: Food wholesaling." *Journal of Food Distribution Research* 6 (2):11–12.

Russell, Harry L.
   1912   Letter From Russell to L. A. Baker. University of Wisconsin, College of Agriculture Administration, General Subject Files, Box 10, December 28. Madison.
   1931   "Commercial support for agricultural research." Pp. 224–27 in *Proceedings of the 45th Annual Convention of the Association of Land-Grant Colleges and Universities.* Chicago.

Ruttan, Vernon W.
   1971   "Research institutions: Questions of organization." Pp. 129–38 in Melvin G. Blase (ed.), *Institutions in Agricultural Development.* Ames: Iowa State University Press.
   1978   "Reviewing agricultural research programmes." *Agricultural Administration* 5:1–19.
   1980   "Bureaucratic productivity: The case of agricultural research." *Public Choice* 35:529–47.
   1982a   *Agricultural Research Policy.* Minneapolis: University of Minnesota Press.
   1982b   "Changing role of public and private sectors in agricultural research." *Science* 216:23–29.

Sailer, Reece
   1977   "On the threshold of the second quarter century." *Bulletin of the Entomological Society of America* 23:1–2.

Salisbury, Glenn W.
   1980   *Research Productivity of The State Agricultural Experiment Station System: Measured by Scientific Publication Output.* Bulletin 762. Urbana: University of Illinois Agricultural Experiment Station.

Salmon, Samuel C., and A. A. Hanson
   1964   *The Principles and Practice of Agricultural Research.* London: L. Hill.

Savage, J. E.
   1975   "Problems and opportunities." *Poultry Science* 54:331–34.

Sawyer, Gordon
1971   *The Agribusiness Poultry Industry.* New York: Exposition Press.

Schiff, Ashley L.
1962   *Fire and Water: Scientific Heresy in the Forest Service.* Cambridge, Massachusetts: Harvard University Press.

Schultz, Theodore W.
1941   *Training and Recruiting of Personnel in the Rural Social Studies.* Washington, D.C.: American Council on Education.

Science
1884   "Comment and Criticism." *Science* 4:507–9.
1885   "Comment and Criticism." *Science* 5:21.

Science Citation Index
1978   *Science Citation Index.* Philadelphia: Institute for Scientific Information.

Scott, C.
1962   "The use of technical literature by industrial technologists." *IEEE Transactions on Engineering Management* EM-9:76–86.

Scott, Roy V.
1970   *The Reluctant Farmer.* Urbana: University of Illinois Press.

Shaw, R. R.
1956   *Pilot Study on the Use of Scientific Literature by Scientists.* Washington, D.C.: National Science Foundation.

Shepard, Herbert A.
1957   "Basic research and the social system of pure science." Pp. 114–28 in Robert T. Livingston and Stanley H. Milberg (eds.), *Human Relations in Industrial Research Management.* New York: Columbia University Press.

Shilling, C. W.; J. Bernard and J. W. Tyson
1964   *Informal Communication Among Bioscientists.* Washington, D.C.: George Washington University.

Snedecor, George W., and William G. Cochran
1980   *Statistical Methods.* 7th Ed. Ames: Iowa State University Press.

Spitz, Pierre
1975   "Notes Sur L'Histoire des Transferts de Techniques dans le Domaine de la Production Vegetale." Paper presented at Seminar on Science, Technology, and Development, Organization for Economic Cooperation and Development. Paris.

Splinter, W. E.
1980   "Agricultural mechanization: Who wins? Who loses?" *Agricultural Engineering* 61 (May):14–17.

Spring, Joel H.
1972   *Education and the Rise of The Corporate State.* Boston: Beacon Press.

Sprott, Michael J.
1978   "Research needs in agricultural economics from the extension perspective: Attitudes vs. an inventory of needs." *Southern Journal of
Agricultural Economics* 10:13–17.

Stewart, Robert E.
1979   *Seven Decades That Changed America.* St. Joseph, Michigan: American
Society of Agricultural Engineers.

Storer, Norman W.
1961   "Science and Scientists in an Agricultural Research Organization:
A Sociological Study." Ph.D. dissertation, Cornell University.

Strauss, Anselm L.
1978   *Negotiations: Varieties, Contexts, Processes, and Social Order.* San Francisco: Jossey-Bass.

Strauss, Anselm L.; L. Schatzman; D. Ehrlich; R. Bucher; and M. Sabshin
1963   "The hospital and its negotiated order." Pp. 147–49 in E. Friedson
(ed.), *The Hospital in Modern Society.* New York: Free Press.

Sullivan, D.; D. H. White; and E. J. Barboni
1977   "The state of a science: Indicators in the specialty of weak interactions." *Social Studies of Science* 7:167–200.

Taylor, Frederick Winslow
1911   *The Principles of Scientific Management.* New York: Harper.

Thompson, C. W.
1917   "Rural surveys." Pp. 129–33 in *Papers and Proceedings, 11th Annual
Meeting, American Sociological Society.* Columbus, Ohio.

Tidbury, G. E.
1974   "Communication between agricultural scientists: Work of the Com-
monwealth's Agricultural Bureaux." *Agricultural Administration*
1:103–14.

Trigo, Eduardo, and Martin E. Piñeiro
1981   "Dynamics of agricultural research organization in Latin America."
Food Policy 6:2–10.

True, Alfred Charles
1929   *History of Agricultural Education in the United States 1785–1925.* USDA,
Miscellaneous Publication No. 36. Washington, D.C.
1937   *A History of Agricultural Experimentation and Research in the United
States.* USDA, Miscellaneous Publication No. 251. Washington, D.C.

Ulrich's International Periodicals Directory
1978   *Ulrich's International Periodicals Directory.* 18th ed. New York: R. R.
Bowker Co.

UNESCO
1981   *1981 Statistical Yearbook.* Paris: United Nations Educational, Scientific,
and Cultural Organization.

U.S. Congress, House of Representatives. Committee on Science and Technology
1976   *Special Oversight Review of Agricultural Research and Development.* Washington, D.C.: USGPO.

United States Department of Agriculture
1978   *Agricultural and Food Research Issues and Priorities, A Review and Assessment.* Report to the Agricultural Research Policy Advisory Committee (ARPAC), the Joint Council, NASULGC and USDA. Washington, D.C: USGPO.
1981   *1981 Handbook of Agricultural Charts.* Agricultural Handbook No. 592. Washington, D.C.: USDA.

USDA, Agricultural Research Service
1967   *The Use of Quality and Quantity of Publication as Criteria for Evaluating Scientists.* USDA, ARS, Miscellaneous Publication No. 1041. Washington, D.C.

USDA, Cooperative State Research Service
1896–   *Funds for Research at State Agricultural Experiment Stations* (Title varies).
1975   Washington, D.C.: USDA, CSRS.

USDA, Current Research Information System
1978   *Manual of Classification of Agricultural and Forestry Research.* Revision III. Washington, D.C.: USDA, CRIS.

USDA, Economic Research Service
1963   *Handbook of Agricultural Charts.* USDA, ERS, Handbook No. 258. Washington, D.C.
1977   *Research Issues Facing Agriculture and Rural America.* Washington, D.C.: USDA, ERS.

USDA, Interbureau Committee on Technology
1940   *Technology on the Farm.* Washington, D.C: USGPO.

USDA, Science and Education Administration
1980   *Proceedings of the Conference on Priorities for Agricultural Research, Extension, and Higher Education,* January 27–28. Arlington, Virginia.
1981   *Inventory of Agricultural Research FY 1978.* Beltsville, Maryland: USDA.

USDA, Special Work Group
1977   *Basic Research in Agriculture. A Report to the National Science Board and the Agricultural Research Policy Advisory Committee.* Washington, D.C.: USGPO.

U.S. Department of Commerce, Bureau of the Census
1978   *Selected Characteristics of Persons in Fields of Science or Engineering, 1976.* Special Studies P-23, No. 76. Washington, D.C.: USGPO.
1979   *Farm Population of the United States: 1978.* Washington, D.C.: USGPO.

U.S. General Accounting Office
1977   *Management of Agricultural Research, Need and Opportunities for Improvement.* Report to the Joint Economic Committee by the Comptroller General of the United States. Washington, D.C.: USGPO.

1978   *Changing Character and Structure of American Agriculture: An Overview.* Washington, D.C: USGPO.

1981   *Cooperative Extension Service's Mission and Federal Role Need Congressional Clarification.* Report to the Congress by the Comptroller General of the United States. Washington, D.C.: USGPO.

U.S. Office of Education
1901   *Report of the Commissioner of Education, 1900.* Washington, D.C.: USGPO.

U.S. Office of Technology Assessment
1978   *Nutrition Research Alternatives.* Washington, D.C.: USGPO.

1981   *An Assessment of the United States Food and Agricultural Research System.* Washington, D.C.: USGPO.

Vacin, Gary L.
1979   "How land-grant universities wholesale information to farm magazines." Manhattan, Kansas: Science and Education Administration-Extension, USDA and Kansas State University.

Vogeler, Ingolf
1981   *The Myth of the Family Farm: Agribusiness Dominance of U.S. Agriculture.* Boulder, Colorado: Westview Press.

Volk, N. J.
1960   "Role of the experiment stations in AALGCSU affairs." Pp. 133–34 in *Proceedings of the 74th Annual Convention of the American Association of Land-Grant Colleges and State Universities.* Washington, D.C.

Volmer, Howard M.
1972   "Basic and applied research." Pp. 67–96 in Saad Nagi and Ronald G. Corwin (eds.), *The Social Contexts of Research.* New York: John Wiley.

von Bertalanffy, Ludwig
1968   *General Systems Theory.* New York: George Braziller.

Walsh, John
1981   "Biotechnology boom reaches agriculture." *Science* 213:1339–41.

Warner, Frederick B., and Robert L. Bruce
1975   "Cornell researchers study 'Publish or Perish' notion." *ACE* 58:(4) (October–December):22–29.

Webster, Bayard
1979   "6 scientists quit panel in dispute over livestock drugs." *The New York Times* (January 23), C2.

Weinstein, Deena
1976   "Determinants of problem choice in scientific research." *Sociological Symposium* 16:13–23.

Wellman, Harry Richard
1977   *Teaching, Research, and Administration: University of California 1925–1968.*

University of California, Agriculture Oral History Collection No. 24. Berkeley.

West, Quentin M.
1973 "Economic research trade-offs between efficiency and equity." *Agricultural Science Review* 11 (First Quarter):31–34.

Whisnant, Susan
1976 "Attitudes faced by female foresters." *Journal of Forestry* 74:762.

White, B.
1970 *Planners and Information.* London: Library Association.

Whitley, Richard
1974 "Cognition and social institutionalization of scientific specialties and research areas." Pp. 69–95 in Richard Whitley (ed.), *Social Processes of Scientific Development.* London: Routledge and Kegan Paul.

Wiest, Edward
1923 *Agricultural Organization in the United States.* Lexington: University of Kentucky Press.

Wittwer, Sylvan H.
1980 "Food production prospects: Technology and resources options." Pp. 60–99 in D. Gale Johnson (ed.), *The Politics of Food.* Chicago: Chicago Council on Foreign Relations.

Woods, Albert F.
1926 "Presidential address—The relation of land-grant institutions to the
[1925] fundamentals of forward looking national policies for the development of agriculture." Pp. 22–29 in *Proceedings of the 39th Annual Convention of the Association of Land-Grant Colleges.* Chicago.

Yang-Boo, Choe
1978 "Toward an idea of agricultural economics: A critique of the idea of the applied economics of agriculture." *Journal of Rural Development* 1:1–21.

Youngberg, Garth
1978 "The alternative agriculture movement." *Policy Studies Journal* 6:524–30.

Zuckerman, Harriet
1978 "Theory choice and problem choice in science." *Sociological Inquiry* 48 (3-4):65–95.

Abbreviations

AAACES	American Association of Agricultural Colleges and Experiment Stations
AAAL	American Association for Agricultural Legislation
AFBF	American Farm Bureau Federation
ARPAC	Agricultural Research Policy Advisory Committee
ARS	Agricultural Research Service
ASAE	American Society of Agricultural Engineers
BAE	Bureau of Agricultural Economics
CAST	Council on Agricultural Science and Technology
CRIS	Current Research Information System
ERS	Economic Research Service
ESA	Entomological Society of America
FSR	Farming Systems Research
GAO	General Accounting Office
IPM	Integrated Pest Management
LGCs	Land-Grant Colleges
NASULGC	National Association of State Universities and Land-Grant Colleges
NIH	National Institutes of Health
NRC	National Research Council
NSF	National Science Foundation
NTIS	National Technical Information Service
OES	Office of Experiment Stations
OTA	Office of Technology Assessment
RPA	Research Problem Area
SAES	State Agricultural Experiment Stations
SEA—AR	Science and Education Administration—Agricultural Research
UNESCO	United Nations Educational, Scientific, and Cultural Organization
USAID	United States Agency for International Development
USDA	United States Department of Agriculture
USDC	United States Department of Commerce
USGPO	United States Government Printing Office

Appendix:
Sources of Information
and Methods

The research reported in this book involved a variety of sources of information. These included a review of historical materials; a series of in-depth interviews with agricultural scientists at several institutions; a review of published documents by various agricultural scientific and disciplinary associations; and two mail surveys, the first a survey of approximately 2,000 agricultural scientists, the second of U.S. agricultural journal editors. These diverse sources yielded complementary information that was later synthesized into this volume.

Historical Review

Historical materials from the agricultural sciences were systematically reviewed for this study. Secondary sources such as Rosenberg (1976) and Danbom (1979) were particularly helpful. In addition, we consulted a number of primary sources, including the annual proceedings of what is now known as the National Association of State Universities and Land-Grant Colleges. Also important were the proceedings of the Country Life Association, the printed papers of the American Association for Agricultural Legislation, and various histories of state colleges of agriculture.

Official Documents

Documents of the various agricultural scientific disciplinary associations, such as the American Society of Agronomy, the Entomological Society of America, and the American Agricultural Economics Association, were reviewed for state-of-the-discipline papers, presidential addresses, and commentary relating to disciplinary directions. Such publications over the last five years were reviewed systematically; earlier documents were reviewed when cited. In addition, we searched official government documents and selected experiment station reports.

Interviews

We conducted in-depth interviews with agricultural scientists at several institutions. Interviewees were selected by a process known as theoretical sampling (Glaser and Strauss, 1967), which attempts to maximize the variance among the respondents. To facilitate this procedure, respondents were asked at the conclusion of each interview to identify another potential interviewee who would have substantially different views on the discipline. During the course of the interviews, the questionnaire was modified. Questions eliciting invariant responses were dropped, and new questions were added that reflected comments made by the respondents. This procedure provided initial insight into the agricultural sciences and insured that the questions asked in the mail questionnaire would be relevant to the respondents.

Mail Surveys

Scientists' Questionnaire (Figure A.1)

The statistical results reported herein were based upon a mail survey of agricultural scientists. Questionnaires were sent to a random sample (n = 2051) of principal investigators listed in the Current Research Information System (CRIS). As the CRIS file contains names of all persons engaged in research at institutions that receive federal agricultural funds, it proved to be an ideal sampling frame for this study.

Employing techniques developed by Dillman et al. (1974), we mailed questionnaires to scientists with a cover letter stating the intentions behind the survey. Several weeks later, we sent a postcard reminder to scientists who had not yet replied. Several weeks after that, we sent a second letter and questionnaire to those who had not yet replied. Finally, a third wave of questionnaires was sent via certified mail.

One hundred seventy-five questionnaires were excluded from the sample because of death, retirement, exit from agricultural research, and other reasons. This left a corrected sample totaling 1,876 scientists. Of these, 1,431 returned their questionnaires, for a 76% response rate. This compares very favorably with questionnaires sent to the general public, where response rates rarely exceed 60%.

Most questions were precoded so data could be transferred directly onto magnetic tape. However, certain questions, such as those relating to educational background and the names of journals in which scientists had published, could not be precoded. These items were first coded onto standard code sheets and then transferred to magnetic tape. Finally, approximately one-third of the respondents made some remarks on the "comments" page of the questionnaire. These remarks were transferred to 4" x 6" cards and used as an aid in interpreting the statistical data. Several of the comments are quoted in the text.

Nonresponse rates to individual questions on the questionnaire were relatively low. In general, at least 1,300 persons responded to each question.

(Text continues on p. 289.)

FIGURE A.1

RESEARCH PROBLEM FORMULATION IN THE AGRICULTURAL SCIENCES

Scientists' Questionnaire

I. In order to analyze research problem choices, we need to first examine the degree to which personal, educational and career background contribute to one's research strategy. Therefore, we would appreciate your answers to the following questions. Thank you for your cooperation.

What is your sex?

___Male (0) ___Female (1)

What is your year of birth? 19___

What is the size of the community where you were living at age 16?

(1)___Open country
(2)___Town of 2,499 or less
(3)___Town of 2,500 - 9,999
(4)___Town of 10,000 - 24,999
(5)___Town of 25,000 - 50,000
(6)___City of more than 50,000

What was the principal occupation of your father when you were age 16 (if deceased or retired, indicate previous occupation)?

(01)___Worker (unskilled or semi-skilled)
(02)___Skilled worker, mechanic or foreman
(03)___Farm worker or tenant
(04)___Farm owner
(05)___Farm manager
(06)___Clerk or retail salesman
(07)___White-collar worker or salesman
(08)___Minor executive
(09)___Major executive
(10)___Owner small business
(11)___Owner medium or large business
(12)___Teacher (specify subject and level)

_____

(98)___Other (specify)_____

What is your educational background?

Degree	Institution	Year Completed	Field or Department
Baccalaureate			
Masters			
Doctorate			
Postdoctorate			
Other (specify)			

Please indicate the title, duties, years on the job and type of setting (academic department, government lab, etc.) for your present, previous and first professional position(s).

	Title	Number of Years Employed	Institution & Department
Present position			
Previous position			
First professional position			

What is your current field of science?_____

-2-

During the last 12 months what have been the *actual* (not formal) conditions of
your appointment?

_____% Research   _____% Administration
_____% Teaching   _____% Extension

_____% Other (specify)_____

II. In this next section we are interested in many aspects of your work, your choice of
research topics and your opinions about your discipline. Some of the questions are
complex and others call for opinions and ideas on topics to which you may not have
given much thought. Please try to answer all of the questions. When the precoded
answers do not reflect your situation or attitudes, please check the open ended
response and elaborate.

The National Science Foundation defines basic science, applied science, and
development in the following ways:

Basic research: stresses that it is directed toward increases of
                        knowledge in science with ". . . the primary aim
                        of the investigator . . . a fuller knowledge or
                        understanding of the subject under study, rather
                        than a practical application thereof."

Applied research: is directed toward practical application of
                        knowledge. It covers ". . . research projects
                        which represent investigations directed to
                        discovery of new scientific knowledge and which
                        have specific commercial objectives with respect
                        to either products or processes."

Development: development may be summarized as ". . .the
                        systematic use of scientific knowledge directed
                        toward the production of useful materials,
                        devices, systems or methods, including design
                        and development of prototypes and processes."

Using these definitions, how would you characterize your research during the last
5 years (1974-1979)? (Indicate percentage of research time.) What do you think
it should be?

Actual %   Ideal %
_____Basic research   _____
_____Applied research _____
_____Development   _____

What is your perception of the distribution of research in your department during
the last 5 years? Ideally, what should it be?

Actual %   Ideal %
_____Basic research   _____
_____Applied research _____
_____Development   _____

-3-

During the last 5 years, how important were the following considerations in your choice of research problems? Please rate *each* criterion by circling one number from *"Not Important"* (1), to *"Very Important"* (7).

CRITERIA FOR PROBLEM CHOICE

	Not Important					Very Important
Potential contribution to scientific theory	1 2 3 4 5 6 7
Likelihood of clear empirical results	1 2 3 4 5 6 7
Potential creation of new methods, useful materials and devices	1 2 3 4 5 6 7
Potential marketability of the final product	1 2 3 4 5 6 7
Funding	1 2 3 4 5 6 7
Length of time required to complete the research	1 2 3 4 5 6 7
Publication probability in professional journals	1 2 3 4 5 6 7
Publication probability in farm and/or industry journals. .	1 2 3 4 5 6 7
Publication probability in experiment station or research service bulletins and reports	1 2 3 4 5 6 7
Availability of research facilities	1 2 3 4 5 6 7
Currently a 'hot' topic	1 2 3 4 5 6 7
Evaluation of research by scientists in your field	1 2 3 4 5 6 7
Colleagues' approval	1 2 3 4 5 6 7
Credibility of other investigators doing similar research .	1 2 3 4 5 6 7
Enjoy doing this kind of research	1 2 3 4 5 6 7
Importance to society	1 2 3 4 5 6 7
Scientific curiosity	1 2 3 4 5 6 7
Demands raised by clientele	1 2 3 4 5 6 7
Feedback from extension personnel	1 2 3 4 5 6 7
Client needs as assessed by you	1 2 3 4 5 6 7
Priorities of the research organization	1 2 3 4 5 6 7
Other (specify)_____	1 2 3 4 5 6 7

-4-

In choosing your research during the last 5 years, in what ways did the following people influence you? For each person circle the appropriate letter(s). Then indicate the *degree* of influence by putting a *number* from 1 to 7 on the line next to each circled letter indicating whether the person was *slightly important* = 1 to *most important* = 7.

A - choice of research problem
B - choice of research method
C - identification of key concepts
D - choice of theoretical orientation

A___ B___ C___ D___ . . Your immediate supervisor

A___ B___ C___ D___ . . A colleague in your department

A___ B___ C___ D___ . . A colleague in another agricultural science department at your institution

A___ B___ C___ D___ . . A colleague in a basic science department at your institution

A___ B___ C___ D___ . . A colleague at another institution

A___ B___ C___ D___ . . A research assistant/technician

A___ B___ C___ D___ . . A graduate or postdoctoral student

A___ B___ C___ D___ . . A former professor

A___ B___ C___ D___ . . Director of your research facility

A___ B___ C___ D___ . . Client or potential user

A___ B___ C___ D___ . . Research review committee

A___ B___ C___ D___ . . Other (specify) _____

How frequently do you communicate with the following people regarding your research? (Please circle one number for *each* group.)

	Rarely	Monthly	Bi-weekly	Weekly	Daily
Scientists in your department	1	2	3	4	5
Agricultural scientists outside your department	1	2	3	4	5
Other (non-agricultural) scientists . .	1	2	3	4	5
Administrators	1	2	3	4	5
Clients	1	2	3	4	5
Funding agencies	1	2	3	4	5
Extension staff	1	2	3	4	5

How many of the following persons are currently working under your direction?

Graduate Students _____
Post-doctoral fellows _____
Technicians _____
Other (specify):_____ _____

How frequently do you see the following people <u>socially</u> (i.e., away from work)?

	Rarely	Monthly	Bi-weekly	Weekly	Daily
Scientists in your department	1	2	3	4	5
Agricultural scientists outside your department	1	2	3	4	5
Other scientists	1	2	3	4	5
Clients or potential users of your research	1	2	3	4	5
Non-scientists (e.g., neighbors, relatives)	1	2	3	4	5

Please indicate by circling the appropriate number(s) of all the relationships you have had with each of the following organizations during the past 5 years.

┌──────────────── Research grant or fellowship as principal investigator
│   or in another capacity
│ ┌────────────── Research contract
│ │ ┌──────────── Member of grant review panel or study group
│ │ │ ┌────────── Member of advisory board or group
│ │ │ │ ┌──────── Regular consultant
│ │ │ │ │ ┌────── Occasional consultant
│ │ │ │ │ │
│ │ │ │ │ │   Federal Agency or Unit

A	B	C	D	E	F	USDA (excluding fund regularly budgeted for your research facility)
A	B	C	D	E	F	Department of Defense
A	B	C	D	E	F	Department of Interior
A	D	C	D	E	F	NASA
A	B	C	D	E	F	National Science Foundation
A	B	C	D	E	F	NIH
A	B	C	D	E	F	Department of State (including AID)
A	B	C	D	E	F	FDA
A	B	C	D	E	F	Other federal unit_____

Other

A	B	C	D	E	F	A foundation
A	B	C	D	E	F	Independent non-profit research organization
A	B	C	D	E	F	Local government agency or unit
A	B	C	D	E	F	State government agency or unit
A	B	C	D	E	F	An international organization (e.g., FAO)
A	B	C	D	E	F	Publishing firm
A	B	C	D	E	F	Commodity association
A	B	C	D	E	F	Private corporation
A	B	C	D	E	F	Other_____

-6-

What has been your average annual *research* expenditure (including all salaries) over the past 5 years, to the nearest $1,000 excluding overhead?

$_____,000

Would you have pursued your current research if the funding had been unavailable?

(1)___yes, other support would have been available
(2)___yes, even without other support
(3)___yes, but on a reduced scale
(4)___yes, but in a substantially different form
(5)___no
(6)___other (specify)_____

For a person 5 to 10 years beyond a Ph.D. degree how important are the following factors in his or her reappointment, promotion, and tenure in your department, institute or center. How important do you personally feel these factors ought to be?

	Current Policy		Ideal Policy	
	NOT IMPORTANT	VERY IMPORTANT	NOT IMPORTANT	VERY IMPORTANT
Publications of <u>many</u> scholarly papers	1 2 3 4 5		1 2 3 4 5	
Publication of <u>high</u> quality scholarly papers	1 2 3 4 5		1 2 3 4 5	
Publication of in-house papers and reports	1 2 3 4 5		1 2 3 4 5	
Consulting for or advisory work with the government or foundation . . .	1 2 3 4 5		1 2 3 4 5	
Consulting for or advisory work with private firms	1 2 3 4 5		1 2 3 4 5	
Receipt of a grant or contract	1 2 3 4 5		1 2 3 4 5	
Teaching evaluations	1 2 3 4 5		1 2 3 4 5	
Extension evaluations	1 2 3 4 5		1 2 3 4 5	
Other (specify)_____	1 2 3 4 5		1 2 3 4 5	

Are there topics that are consistently avoided in your field? (Circle one number.)

1 Yes
0 No

If so, which ones?_____

_____

Why?_____

_____

-7-

Over the last 5 years, <u>how many</u> of each of the following types of publications have you authored or co-authored?

<u>Authored</u>   <u>Co-authored</u>

_____Journal articles_____

_____Books . . . . . _____

_____Book chapters . _____

_____Abstracts . . . _____

_____Bulletins . . . _____

_____Reports . . . . _____

_____Other . . . . . _____

To what journals do you subscribe?

a._____   f._____

b._____   g._____

c._____   h._____

d._____   i._____

e._____   j._____

Over the past 5 years, in what journals have you published (authored or co-authored) articles?

a._____   f._____

b._____   g._____

c._____   h._____

d._____   i._____

e._____   j._____

During the last 12 months, about what percent of your <u>research</u> time did you spend:

in the office?   _____%
in the library?   _____%
in the field?   _____%
in the laboratory?   _____%
in the computing facility?   _____%
in the greenhouse?   _____%
elsewhere?
(specify)_____%

The following statements pertain to various criteria that might be relevant to scientific editors (in your field) in their decision as to whether or not to publish a paper. Would you please rate each criterion, based upon how important you believe it to be by circling one number from "*Not Important*" (1), to "*Very Important*" (7)?

SCIENTIFIC EDITORS' CRITERIA	NOT IMPORTANT ▼					VERY IMPORTANT ▼	
The theoretical relevance of the question investigated	1	2	3	4	5	6	7
The presence of original empirical evidence	1	2	3	4	5	6	7
The scholarship demonstrated in the article	1	2	3	4	5	6	7
The relevance of the article to the journal's focus	1	2	3	4	5	6	7
The background and reputation of the author	1	2	3	4	5	6	7
The creativity of ideas in the article	1	2	3	4	5	6	7
The entertainment quality of the essay	1	2	3	4	5	6	7
The ethical sense demonstrated by the author	1	2	3	4	5	6	7
The value of the author's findings to the advancement of the field	1	2	3	4	5	6	7
The sophistication of the author's research methodology and data analysis	1	2	3	4	5	6	7
The potential contribution of the article to increased agricultural productivity	1	2	3	4	5	6	7
The grasp of the author's research design on the question investigated	1	2	3	4	5	6	7
The value of the article's findings to clientele	1	2	3	4	5	6	7
The current popularity of the topic	1	2	3	4	5	6	7
The background and reputation of the author's institution	1	2	3	4	5	6	7
The degree to which an article addresses a controversial issue in the field	1	2	3	4	5	6	7

How important are the following resources in your research?	NOT IMPORTANT ▼					VERY IMPORTANT ▼	
Domestic journals in your field	1	2	3	4	5	6	7
Domestic journals in related fields.	1	2	3	4	5	6	7
Foreign journals in your field	1	2	3	4	5	6	7
Foreign journals in related fields	1	2	3	4	5	6	7
Books and monographs	1	2	3	4	5	6	7
Research bulletins	1	2	3	4	5	6	7

-9-

Below is a list of eleven goals for agricultural research often used by USDA and other agencies. These goals vary in importance and in the degree to which a given research project reflects any or all of them. We would appreciate it if you would tell us how important you believe each goal to be *and* the degree to which your research contributes to each goal.

GOALS

Importance to you as a Goal		Degree Your Research Contributes to this Goal
OF NO IMPORTANCE — OF HIGHEST IMPORTANCE		OF NO IMPORTANCE — OF HIGHEST IMPORTANCE

1 2 3 4 5 6 7......Increase agricultural
productivity...............1 2 3 4 5 6 7

Protect forests, crops,
and livestock from insects,
1 2 3 4 5 6 7......diseases and other hazards.1 2 3 4 5 6 7

Decrease production costs
1 2 3 4 5 6 7......of farm/forest products....1 2 3 4 5 6 7

Expand demand by developing
new products or enhancing
1 2 3 4 5 6 7......product quality............1 2 3 4 5 6 7

Improve marketing
1 2 3 4 5 6 7......efficiency.................1 2 3 4 5 6 7

Expand export
1 2 3 4 5 6 7......markets....................1 2 3 4 5 6 7

Assist developing
1 2 3 4 5 6 7......nations....................1 2 3 4 5 6 7

Protect consumer health
1 2 3 4 5 6 7......and improve nutrition......1 2 3 4 5 6 7

Improve level of living
1 2 3 4 5 6 7......of rural America...........1 2 3 4 5 6 7

Promote community
1 2 3 4 5 6 7......improvement................1 2 3 4 5 6 7

Develop new knowledge or
1 2 3 4 5 6 7......improved methodology.......1 2 3 4 5 6 7

Apart from your discipline, do you believe that your research and publishing over the past 5 years has already or will directly or indirectly benefit any of the following? In your opinion who should your research benefit?

Will or Does Benefit						Should Benefit				
Not at All →			A Great Deal →			Not at All →			A Great Deal →	

				Other scientific disciplines					
1 2 3 4 5	.......(specify)_____	.......1 2 3 4 5
1 2 3 4 5	......Small farmers.....................1 2 3 4 5
1 2 3 4 5	......Large farmers.....................1 2 3 4 5
1 2 3 4 5	......Agri-business.....................1 2 3 4 5
1 2 3 4 5	......Rural residents...................1 2 3 4 5
1 2 3 4 5	......General public....................1 2 3 4 5

1 2 3 4 5......Local or state governmental agencies...........................1 2 3 4 5

1 2 3 4 5......Federal agencies..................1 2 3 4 5

1 2 3 4 5......Foreign groups, institutions or governments....................1 2 3 4 5

1 2 3 4 5......Other_____.......1 2 3 4 5

Recently, a number of social issues relating to agriculture have been raised by various groups and individuals. Would you please indicate the degree to which your research reflects these concerns? (Please circle one number for *each* concern.)

Rarely → ... Frequently →

Consumer issues (e.g., nutrition). 1 2 3 4 5 6 7

Energy issues (e.g., fossil fuel usage, energy efficiency) .1 2 3 4 5 6 7

Environmental issues (e.g., farm runoff, pest management). .1 2 3 4 5 6 7

Alternative approaches to agriculture (e.g., organic farming) 1 2 3 4 5 6 7

World Food "Crisis" (e.g., problems of food sufficiency and distribution) 1 2 3 4 5 6 7

Other (specify)_____ 1 2 3 4 5 6 7

-11-

How would you characterize your research orientation, "school" or paradigm within your discipline (e.g., until recently some physicists treated light as a particle while others treated it as a wave)?

_____

_____

_____

About what percent of the members of the discipline share your orientation?

_____%

What are the current "hot" specialties within your discipline?

_____

_____

_____

Why are they "hot"?

_____

_____

_____

Are there any research interests you would like to pursue which you have not been able to up until now?

_____Yes (1)

_____No (0)

If so, what would make it possible for you to do these things?

_____

_____

_____

No questionnaire of this type can adequately cover points considered relevant by individuals with diverse interests. In light of this, we would appreciate your providing any additional comments that you may have on the next page.

THANK YOU FOR YOUR COOPERATION

288

COMMENTS _____

_____

_____

_____

_____

_____

_____

_____

_____

_____

_____

_____

_____

_____

_____

_____

_____

_____

_____

_____

_____

_____

_____

_____

_____

_____

_____

_____

_____

_____

_____

_____

_____

The College of Agriculture is an Equal Opportunity Organization authorized to provide research, educational information and other services only to individuals and institutions that function without regard to race, color, sex, age, handicap or national origin.

One exception was the question dealing with age, which elicited an 80% response rate. This appears to have been due to its placement on the questionnaire. As the response rates do not vary greatly, we have not reported them individually.

For statistical analysis, we used the Statistical Package for the Social Sciences and the Statistical Analysis System whenever possible. In certain cases involving particularly complex data manipulation, special programs were written.

Journal Editors' Questionnaire

We determined the population of agricultural-scientific journal editors after a careful review of the journals. Our journal list was compiled from all journals (1) indexed by the agricultural library at the authors' university, (2) listed under agricultural subject headings in the 1978 *Science Citation Index*, and (3) arranged under agricultural subject headings in *Ulrich's International Periodicals Directory* (18th ed.). Foreign journals, newsletters, and industry/trade journals were not included in the final list of 103 journals. Several journals were difficult to classify because they included nonresearch articles. However, we included them because they also published refereed original research. Virtually all United States–based agricultural-scientific journals were included.

To collect data from the editors, we used a self-administered mail questionnaire with techniques similar to those used in the survey of scientists. After the fourth (certified) mailing, 92 editors (89%) returned completed questionnaires. For more information on the journals and journal editors, see Lacy and Busch (1982b).

Statistical Analysis of Quantitative Data

Nearly all of the quantitative data generated through our mail questionnaires were statistically analyzed by means of analysis of variance or Pearsonian correlations. We chose these statistics largely because they are used by both social scientists and agricultural scientists. For similar reasons, we avoided scales and nonparametric statistics.

Levels of statistical significance (utilizing student's t or F, as appropriate) are reported in the text and tables. We have usually excluded nonsignificant differences in order to avoid overly complex tables.

In some cases, we used Least Significant Difference (LSD) tests. Although commonly used in the biological sciences, these statistics are rarely used in the social sciences. They are essentially a student's t test between all possible pairs of group means. They make it possible to discern when pairs of group means are statistically different from each other at a given probability level. For example, in Table 2.1, items 1 and 2 are different from each other ($p < 0.001$), as they have different letters (n and o) attached to them. In contrast, items 3 and 4 are not statistically different from each other, as they are both followed by the letter p.

One problem with the LSD test is that it may erroneously find significant differences where none in fact exist. To avoid this problem, we have not used the LSD test in instances where the F is not significant at the 0.01 level. For a review of the use of the statistic, see Snedecor and Cochran (1980).

Definition of Fields of Science

We defined the fields of science from written reports of scientists on the questionnaire. Over 200 fields were initially identified. Some were merely different ways of reporting the same information (e.g., crop science and agronomy-crops). In other cases, respondents identified subfields of particular disciplines (e.g., beef cattle nutrition). In those cases, we combined subfields into larger disciplines (e.g., animal science).

Any classification system of this type must retain a certain level of arbitrariness. Of course, we attempted to obtain the best fit possible. A small number of veterinary, poultry, and dairy scientists were grouped in "animal scientists." Scientists in chemistry, biology, biochemistry, and genetics were grouped together as "basic scientists," as we observed little difference among them on most items. Rural sociologists, home economists, political scientists, and agricultural educators were grouped under "social scientists." "Environmental scientists" included wildlife ecologists, environmental toxicologists, fisheries biologists, and wildlife management specialists. Approximately 50 scientists were unclassifiable and have been excluded from analyses involving breakdowns by discipline.

One interesting note is that foresters and nutritionists frequently commented in the margins of the questionnaire that they were not "agricultural scientists" at all. In our study, of course, we included them under that rubric.

Prestige Scores

The prestige scores used herein were generated by Ladd and Lipset (1979) as part of a national survey of college and university faculty. Specifically, the scores indicate the percentage of respondents in the agricultural sciences who identified a particular institution as the best in their discipline. As these scores were aggregated across the agricultural sciences, they represented gross prestige differences among colleges of agriculture rather than differences within each discipline. In particular, they poorly describe institutions with forestry programs because several forestry departments or colleges are located at separate institutions from those having other agricultural disciplines.

Substantial criticism has been made of the methods used in the Ladd and Lipset study. For a review, see Lang (1981). Without question, the prestige scores are limited by the absence of certain institutions from the sample. Nevertheless, these scores do provide the best recent estimates of institutional prestige available. They also order institutions in much the same way in which a panel of distinguished animal scientists ranked them (Salisbury, 1980:23).

TABLE A.1
USDA Goals of Agricultural Research

I. Insure a stable and productive agriculture for the future through wise management of natural resources.

II. Protect forests, crops, and livestock from insects, diseases, and other hazards.

III. Produce an adequate supply of farm and forest products at decreasing real production costs.

IV. Expand the demand for farm and forest products by developing new and improved products and processes and by enhancing product quality.

V. Improve efficiency in the marketing system.

VI. Expand export markets and assist developing nations.

VII. Protect consumer health and improve nutrition and well-being of the American people.

VIII. Assist rural Americans to improve their level of living.

IX. Promote community improvement including development of beauty, recreation, environment, economic opportunity, and public services.

X. Enhance the national capacity to develop and disseminate new knowledge and new or improved methodology for solving current problems or new problems that will arise in the future. Research under Goal X is conducted under all the RPAs [Research Problem Areas] of Goals I - IX.

Source: USDA, Current Research Information System, Manual of Classification of Agricultural and Forestry Research, Washington: CRIS, 1970, Revision III, 1978.

Goals of Agricultural Research

The USDA employs a list of ten research goals. These are reproduced in Table A.1. The list as a whole is a laudable one, but several major flaws made it unacceptable for our study. First, several individual goals actually consist of more than one goal. For example, goal VI includes both the expansion of export markets and the assistance of developing nations. In addition, goal X is not used as a separate goal for accounting purposes but, rather, is folded into each of the other nine goals.

The USDA reporting system is also problematic in that scientists do not report their personal assessment of the degree to which their research contributes to a given goal. Instead, they are asked to categorize the percent of their research that contributes to a particular "research problem area" (RPA). These RPAs are then grouped under particular goals. Thus, for example, all research on the economic and legal problems of water utilization is classified under goal number I, while wildlife and fish ecology is classified under goal number IX. As a result of these inconsistencies, we adopted a modified, and we hope less ambiguous, version of the USDA list of goals. Scientists then directly estimated their degree of contribution and rated the intrinsic importance for each goal. This list is reported in the text.

Errors and Omissions

This study, like others of this nature, is not free of errors. Most of the errors of which we are aware are in the mail questionnaire. Happily, most errors are also typographical and had little or no observable effect upon responses. However, certain errors of substance did occur. First, as we noted above, the placement of the question on age on the questionnaire apparently resulted in its being overlooked by some respondents. Second, farm workers were excluded from the list of beneficiaries of research. Finally, several open-ended questions that were satisfactorily answered in interviews and pretests were misunderstood or left blank by most respondents. Some issues that might have been addressed in the survey were deliberately omitted because of lack of space in what was an already lengthy survey instrument. We make no apology for these errors and omissions, but simply mention them for the reader's information.

Index